THE YACHTSMAN'S GUIDE TO

celestial navigation

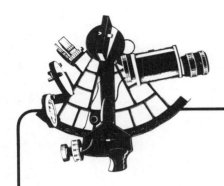

THE YACHTSMAN'S GUIDE TO
celestial navigation
by Stafford Campbell

YACHTING/BOATING BOOKS • Ziff-Davis Publishing Co.
New York

Contents

ABBREVIATIONS COMMONLY USED IN
CELESTIAL NAVIGATION

a	Altitude difference, or Intercept
aλ	Assumed Longitude
aL	Assumed Latitude
AP	Assumed Position
corr	Correction to a tabulated value
d	Declination hourly change, or tabular altitude differential
D	Dip; correction for observer's height of eye
Dec (dec)	Declination of a celestial body
DR	Dead Reckoning (position)
GHA (gha)	Greenwich Hour Angle
GMT	Greenwich Mean Time
GP	Geographical Position (of a celestial body on Earth)
ha (App. Alt.)	Apparent Altitude
hs	Sextant Altitude
Hc	Computed Altitude
Ho	Observed Altitude
H.P.	Horizontal Parallax
IC	Index Correction
incr	Incremental value
LAN	Local Apparent Noon
LHA	Local Hour Angle
Mer Pass	Meridian Passage (Time of)
R	Refraction corrections
SHA	Sidereal Hour Angle
Tab	Tabular value
v	Small, variable additional corrections
W	Watch Time
z	Zenith Distance
Z	Uncorrected Azimuth Angle
Zn	True Azimuth
♈	First Point of Aries (Vernal Equinox)

Foreword

In an electronic age one may wonder why another book on celestial navigation. Particularly one which promotes the old-fashioned use of sextant and chronometer rather than a calculator in a black box.

The answer is to be found with the author. Of all the small-boat navigators I have known Staff Campbell probably enjoys his subject far more than most. This helps account for his tremendous expertise. It explains too, his desire to instill in other amateur sailors his own love and enthusiasm for this art and science.

He comes well qualified, from service aboard a World War II Destroyer in the Pacific to many yacht cruises in Europe, the Mediterranean, and the Caribbean. Most important, he has treated a sobering subject with simplicity and grace.

The past 25 years have produced a generation of young people who are bent on self-improvement through self-taught skills. Many of them yearn for the good old days they have never known. This little book is for them as well as for any sailor who enjoys the intellectual satisfaction of knowing he can find his position on the oceans using the same tools, sighting the same stars and planets, as all the famous navigators of history.

The system taught here (Pub. No. 249) remains the fastest and simplest available today. Its accuracy is wholly compatible with observations possible from the moving deck of a small yacht.

The book was conceived to be used on board; was designed to fit into a pocket convenient for reference. It is not an exhaustive treatise—just the most concise and practical volume for yachts of all sizes.

Sydney H. Rogers
Darien, Connecticut

Introduction

Welcome to the world of celestial navigation. Such a welcome may be a little limited since there are many worlds and satellites of celestial—much like our universe—without finite end. Coming back to earth for a moment, what I really want to welcome you to are the elements of celestial navigation which are of immediate, practical use to the yachtsman.

We'll start with the assumption that you already have a basic familiarity with the fundamentals of piloting, but you've been wondering what celestial is all about and whether you could add it to your kit. Stop wondering and start reading, because celestial is not only the ultimate in the sailor's art but, with modern methods, is not that hard to learn. Being self-contained, and dependent only on the navigator himself, celestial navigation may well bring you home when all else fails.

You will need some equipment in addition to the

charts, protractor and dividers you're familiar with in piloting; namely a sextant, a timepiece, an almanac and a sight reduction table. More about these as we go along.

First let's see if we can't remove some of the mystique that traditionally has clouded the art and, on the principle that you don't have to understand the theory of the internal combustion engine to drive a car, concentrate on the practice at the outset and take a brief look at the theory later. While I may offend some oldtimers by shortcutting here and there, my purpose is to give you as simply and as quickly as possible the capability of taking and working the sights that you, as a practicing yachtsman-navigator, will encounter. Should you become fascinated with the subject, and I warn you you might, there is a lifetime of learning possible and the classic texts—"Bowditch", *American Practical Navigator,* and "Dutton", *Dutton's Navigation and Piloting*— contain it all.

In piloting you learned, besides plotting your course and keeping track of your dead reckoning position (DR) by applying the courses and distances you have sailed from your last known position, how to take bearings on fixed objects such as lighthouses or prominent landmarks ashore. You learned to plot these bearings as lines of position on your chart and, you will recall, the intersection of two or more of these lines gave you a "fix." In Figure I-1, an observer is shown taking bearings of two terrestrial objects, a lighthouse behind a gong buoy bearing 312° True, and a prominent house ashore in range with a nun buoy bearing 036° True.

You will note that the "spread" between the two bearings the observer selected is close to the ideal right angle, thus minimizing any position error resulting

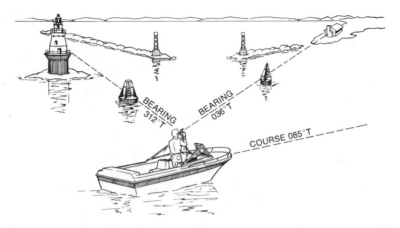

Figure I–1. Observer taking bearings of terrestrial objects with handbearing compass.

from small errors in taking or plotting the bearings. It is generally recommended not to use position lines which intersect at less than 30° unless no better lines are available.

Plotting these two simultaneous bearings on his chart, Figure I-2, the navigator determines his location at the time of observation at the intersection of the position lines—his "fix"—and proceeds from there.

You will also recall that a bearing obtained by radio direction finder, radar or other electronic means similarly produces a line of position which can be used in the same manner, and in conjunction with lines you have obtained visually. By the same token, a distance-off measurement obtained by radar, rangefinder or ver-

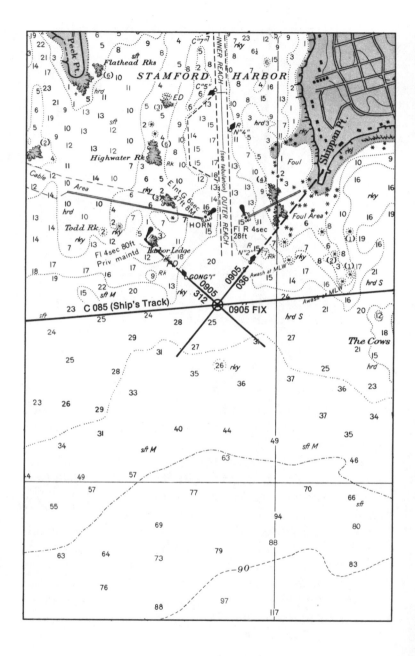

tical sextant angle produces another line of position; in this case a part of the circumference of a circle whose radius is the distance-off.

The key concept is that of the *line of position* which is defined in Bowditch as "a line on some point of which a vessel may be presumed to be located as a result of observation or measurement." You are well on your way to taking some of the mystery out of celestial navigation when you realize that the result of a celestial observation is simply another line of position, and the lines obtained by this means serve exactly the same purpose, and can also be used in conjunction with visually or electronically obtained lines, or any combination of them, in the method you already know.

Leaving the theory until later, let's talk about the six steps you need to follow from the time you choose the celestial body you will shoot to plotting the resulting line of position on your chart. The steps consist of the following:

- The sextant observation
- Timing the sight
- Data from the Almanac
- Calculating the Computed Altitude
- Determining Altitude Difference and Azimuth
- Plotting the line of position

In celestial navigation, as in any human activity from piano-tuning to lion-taming, the practitioners have built up, over the years, a vocabulary which is often

Figure I-2. A fix plotted from two intersecting lines of position

unintelligible to the layman. But don't despair. I'll try to give you a working definition of some of the widely used terms as we go along. For ready reference, at the front of the book is a list of the most commonly used abbreviations, all of which appear in this text, and, in the Glossary at the back of the book, complete explanations of all of the technical nomenclature.

1. The Sextant Observation: The Sun

The sextant observation is the one part of the business in which the operator's technique plays an essential role in its success. It has been said that the rest of modern celestial is "telephone book arithmetic," which I will leave to you to judge, but taking sights from the bouncing deck of a small boat is a question of practice and perseverence, with an ample helping of each.

Much has been written on the subtleties of selecting a sextant, which is beyond the scope of our study. I have achieved good results with a plastic, lifeboat sextant as well as with a brass instrument made in England in the mid-1800's for the clipper ship trade. I've navigated half-way around the world with a Navy Mark 2 from World War II, and sailed the other half with a modern Plath that practically talks. You can get a grounding of navigational fundamentals with almost any workable instrument.

With sextants, as in so many things in life, class tells, and class seems to be closely related to price. The personal equation being equal, the better optics and greater mechanical precision of the more expensive sextants will produce more consistently accurate results. This may be essential to the professional but, and this is an important "but," acceptable results, for the beginner or for the yachtsman who uses celestial but occasionally, can be had with almost any good, well-adjusted instrument. Far larger errors than those inherent in a modestly priced sextant can and do occur in the process of learning the art. Even among experienced navigators, using the best instruments available, it has been found that in a long series of observations, the second thousand sights were more accurate than the first, and further improvement could still be made in the third thousand.

My advice is to start with a sextant you can afford or borrow one which has no more than a minor error. Develop your skill through practice before you invest in one of the super-dupers. A sextant is a delicate, though not fragile, instrument and is worthy of tender, loving care. Keep it clean and free from knocks by returning it to its case after every use and stowing the case in a safe place—not always the easiest accomplishment on a small boat.

The object of the game is to get as accurate as possible

an altitude (the angular distance above the horizon) of the body you are observing with your sextant. This is called the "Sextant Altitude" and is abbreviated hs. To this reading, corrections must be applied for such things as small errors in the sextant, the fact that the height from which you make the observation slightly changes the angle to the horizon and for the bending of the light rays as they come to you through the atmosphere. Modern methods make the determination of all these corrections as easy as looking in a column of numbers.

Since the Sextant Altitude (hs) is the key, let's start there. To begin, we'll try the sun—the body which is most often available and the one you will certainly use most on a small vessel. Holding the sextant (Figure 1–1) comfortably in your right hand, look through the eyepiece at the horizon through the clear half of the horizon glass. If there is too much glare, such as you get on a calm day when the sun is low, use one of the horizon shades. With an appropriate index shade in place—this is very important since the intensity of the unshaded sun's reflection in the eyepiece can injure the eye— move the index arm with your left hand until the image of the sun appears in the mirrored part of the horizon glass. It will actually appear to you that you have brought the sun down to the horizon. Move your left hand to the micrometer drum, or other means of fine adjustment, and proceed to bring the bottom edge, or "lower limb," of the sun so it is just touching the horizon. You will find it takes less time to do than to read about, but the careful navigator will take a second or two more to rotate the sextant back and forth around the line of sight to make sure the image is at the absolute bottom of its arc, and so he knows he has held the instrument in a correct, vertical position.

Quite obviously, both the sun and the horizon must be visible at the same time to take a sight with a marine sextant, but a small boat's horizon is not very far away and you'll be surprised at the hazy conditions in which an experienced navigator can get respectable results.

Figure 1-1. The principal parts of a modern marine sextant

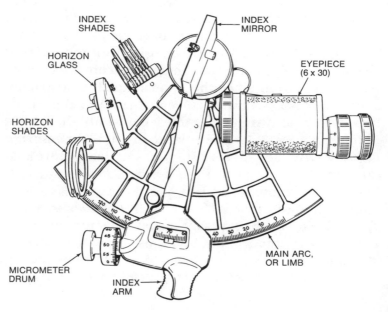

INDEX SHADES · INDEX MIRROR · HORIZON GLASS · EYEPIECE (6 x 30) · HORIZON SHADES · MICROMETER DRUM · INDEX ARM · MAIN ARC, OR LIMB

Note that we have chosen to shoot the lower limb of the sun for our first sight, and this is probably the single sight you will be taking more often than all others combined. You could have used the upper limb had you wished and, as you progress, you will want to try all the other navigational bodies, the particular requirements of which will be covered later.

However, I have a confession to make. After 40 years of taking sights from vessels large and small, I have found that I can get really reliable results from observations of the stars and planets only when I have a good, steady platform—not always possible on a small boat. Despite shattering romantic notions, my daytime observations of the sun from small boats are almost always of a higher degree of accuracy and, normally, quite adequate for the needs of navigation offshore—especially considering the relatively slow speed, shallow draft, and high maneuverability of modern yachts. Thus, I tend to work mostly with the sun, calling on the stars in ideal conditions or, if there has been no opportunity for sun sights, when a landfall is imminent or there are dangers to be avoided. You shouldn't be discouraged from practicing with the other heavenly bodies—in fact, your technique will profit by it—but I recommend highly that you start with the sun and learn it well, as it will be your constant companion.

I'll be surprised if you have any special difficulty in reading your sextant. To avoid a careless error, it is a good idea to read the whole degrees first from the main arc, and then the minutes and any fraction from the micrometer drum or vernier. The sextant illustrated in Figure 1-2, for example, reads 69° 48.5′, which the navigator enters simply as 69-48.5.

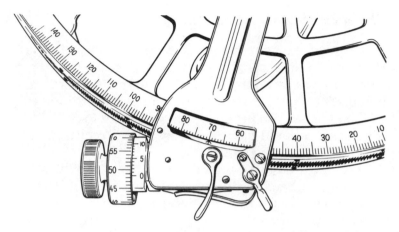

Figure 1-2. Reading the Sextant. Degrees are read from the main arc, or limb; minutes and tenths from the micrometer drum or vernier. The reading illustrated is 69° 48.5'.

Having taken your sun sight and determined Sextant Altitude (hs), it is now time to correct it to obtain the so-called Observed Altitude (Ho) which will be used in the final calculation to derive your line of position. There are three corrections you need to apply to your sun's altitude: Index Correction, Dip and Refraction. These are determined in the first instance by a check of your instrument and, for the latter two, by inspection of the table in the inside front cover of the almanac, a specimen of which greatly reduced, is illustrated in Figure 1-3.

The Index Correction (IC) is necessary because even the finest sextant cannot be expected to stay in perfect adjustment. It is determined before or after taking the sight by setting the index arm at zero degrees and aiming the sextant at the horizon. The line of the horizon will probably appear to be slightly offset between the clear and reflecting halves of the horizon glass. With the fine adjustment, the horizon line is then brought back into coincidence and the sextant read. The correction needed to bring that reading back to zero is the Index Correction (IC). It is usually the practice to check the IC each time the sextant is used, but one check will suffice for a simultaneous series of sights.

For our exercise, let's say that you checked your sextant just before shooting the sun and found that with the horizon line exactly straight the instrument read 1.4 minutes, not zero. Then the IC is *minus* 1.4', the amount necessary to bring the reading back to zero.

The second correction is for Dip (D). This is necessary to compensate for the difference in the angle between the line of sight to the horizon from where you observe it, and the true horizontal. The value of the D correction is taken from the DIP column in the almanac table, an excerpt of which is shown in Figure 1-4. This type of table is called a "critical table"; that is, a table in which a single value is tabulated for all readings between limiting entry values.

The navigator simply enters the table with his height of eye—the height of the sextant's eyepiece above the surface of the water when the sight was taken—and takes out the value for the correction, which, you should note, is always *negative*. Let's say the height of eye for your sun sight was nine feet. The D correction would be

OCT.—MAR. SUN APR.—SEPT.					STARS AND PLANETS		DIP				
App. Alt.	Lower Limb	Upper Limb	App. Alt.	Lower Limb	Upper Limb	App. Alt.	Corr^n	App. Alt.	Additional Corr^n	Ht. of Eye	Corr^n	Ht. of Eye	Corr^n	Ht. of Eye	Corr^n

SUN OCT.—MAR.	SUN APR.—SEPT.	STARS/PLANETS	PLANETS Additional	DIP (m)	DIP (ft)	DIP (m)
9 34 +10·8 −21·5	9 39 +10·6 −21·2	9 56 −5·3	1978	2·4 −2·8 8·0	1·0 − 1·8	
9 45 +10·9 −21·4	9 51 +10·7 −21·1	10 08 −5·2	VENUS	2·6 −2·9 8·6	1·5 − 2·2	
9 56 +11·0 −21·3	10 03 +10·8 −21·0	10 20 −5·1	Jan. 1–July 20	2·8 −3·0 9·2	2·0 − 2·5	
10 08 +11·1 −21·2	10 15 +10·9 −20·9	10 33 −5·0	0 + 0·1	3·0 −3·1 9·8	2·5 − 2·8	
10 21 +11·2 −21·1	10 27 +11·0 −20·8	10 46 −4·9	42 + 0·1	3·2 −3·2 10·5	3·0 − 3·0	
10 34 +11·3 −21·0	10 40 +11·1 −20·7	11 00 −4·9	July 21–Sept. 2	3·4 −3·3 11·2	See table	
10 47 +11·4 −20·9	10 54 +11·2 −20·6	11 14 −4·8	0 + 0·2	3·6 −3·4 11·9	←	
11 01 +11·5 −20·8	11 08 +11·3 −20·5	11 29 −4·7	47 + 0·2	3·8 −3·5 12·6	m	
11 15 +11·6 −20·7	11 23 +11·4 −20·4	11 45 −4·6	Sept. 3–Sept. 29	4·0 −3·6 13·3	20 − 7·9	
11 30 +11·7 −20·6	11 38 +11·5 −20·3	12 01 −4·5	0 + 0·3	4·3 −3·7 14·1	22 − 8·3	
11 46 +11·8 −20·5	11 54 +11·6 −20·2	12 18 −4·4	46 + 0·3	4·5 −3·8 14·9	24 − 8·6	
12 02 +11·9 −20·4	12 10 +11·7 −20·1	12 35 −4·3	Sept. 30–Oct. 14	4·7 −3·9 15·7	26 − 9·0	
12 19 +12·0 −20·3	12 28 +11·8 −20·0	12 54 −4·2	0 + 0·4	5·0 −4·0 16·5	28 − 9·3	
12 37 +12·1 −20·2	12 46 +11·9 −19·9	13 13 −4·1	11 + 0·4	5·2 −4·1 17·4		
12 55 +12·2 −20·1	13 05 +12·0 −19·8	13 33 −4·0	41 + 0·5	5·5 −4·2 18·3	30 − 9·6	
13 14 +12·3 −20·0	13 24 +12·1 −19·7	13 54 −3·9	Oct. 15–Oct. 22	5·8 −4·3 19·1	32 −10·0	
13 35 +12·4 −19·9	13 45 +12·2 −19·6	14 16 −3·8	0 + 0·5	6·1 −4·4 20·1	34 −10·3	
13 56 +12·5 −19·8	14 07 +12·3 −19·5	14 40 −3·7	6 + 0·5	6·3 −4·5 21·0	36 −10·6	
14 18 +12·6 −19·7	14 30 +12·4 −19·4	15 04 −3·6	20 + 0·6	6·6 −4·6 22·0	38 −10·8	
14 42 +12·7 −19·6	14 54 +12·5 −19·3	15 30 −3·5	31 + 0·7	6·9 −4·7 22·9		
15 06 +12·8 −19·5	15 19 +12·6 −19·2	15 57 −3·4	Oct. 23–Nov. 25	7·2 −4·8 23·9	40 −11·1	
15 32 +12·9 −19·4	15 46 +12·7 −19·1	16 26 −3·3	0 + 0·6	7·5 −4·9 24·9	42 −11·4	
15 59 +13·0 −19·3	16 14 +12·8 −19·0	16 56 −3·2	4 + 0·6	7·9 −5·0 26·0	44 −11·7	
16 28 +13·1 −19·2	16 44 +12·9 −18·9	17 28 −3·1	12 + 0·7	8·2 −5·1 27·1	46 −11·9	
16 59 +13·2 −19·1	17 15 +13·0 −18·8	18 02 −3·0	22 + 0·8	8·5 −5·2 28·1	48 −12·2	
17 32 +13·3 −19·0	17 48 +13·1 −18·7	18 38 −2·9		8·8 −5·3 29·2		
18 06 +13·4 −18·9	18 24 +13·2 −18·6	19 17 −2·8	Nov. 26–Dec. 3	9·2 −5·4 30·4	ft.	
18 42 +13·5 −18·8	19 01 +13·3 −18·5	19 58 −2·7	0 + 0·5	9·5 −5·5 31·5	2 − 1·4	
19 21 +13·6 −18·7	19 42 +13·4 −18·4	20 42 −2·6	6 + 0·5	9·9 −5·6 32·7	4 − 1·9	
20 03 +13·7 −18·6	20 25 +13·5 −18·3	21 28 −2·5	20 + 0·6	10·3 −5·7 33·9	6 − 2·4	
20 48 +13·8 −18·5	21 11 +13·6 −18·2	22 19 −2·4	31 + 0·7	10·6 −5·8 35·1	8 − 2·7	
21 35 +13·9 −18·4	22 00 +13·7 −18·1	23 13 −2·3		11·0 −5·9 36·3	10 − 3·1	
22 26 +14·0 −18·3	22 54 +13·8 −18·0	24 11 −2·2	Dec. 4–Dec. 19	11·4 −5·9 37·6	See table	
23 22 +14·1 −18·2	23 51 +13·9 −17·9	25 14 −2·1	0 + 0·4	11·8 −6·0 38·9	←	
24 21 +14·2 −18·1	24 53 +14·0 −17·8	26 22 −2·0	11 + 0·4	12·2 −6·1 40·1	ft.	
25 26 +14·3 −18·0	26 00 +14·1 −17·7	27 36 −1·9	41 + 0·5	12·6 −6·2 41·5	70 − 8·1	
26 36 +14·4 −17·9	27 13 +14·2 −17·6	28 56 −1·8	Dec. 20–Dec 31	13·0 −6·3 42·8	75 − 8·4	
27 52 +14·5 −17·8	28 33 +14·3 −17·5	30 24 −1·7	0 + 0·3	13·4 −6·4 44·2	80 − 8·7	
29 15 +14·6 −17·7	30 00 +14·4 −17·4	32 00 −1·6	46 + 0·3	13·8 −6·5 45·5	85 − 8·9	
30 46 +14·7 −17·6	31 35 +14·5 −17·3	33 45 −1·5		14·2 −6·6 46·9	90 − 9·2	
32 26 +14·8 −17·5	33 20 +14·6 −17·2	35 40 −1·4	MARS	14·7 −6·7 48·4	95 − 9·5	
34 17 +14·9 −17·4	35 17 +14·7 −17·1	37 48 −1·3	Jan. 1–Mar. 22	15·1 −6·8 49·8		
36 20 +15·0 −17·3	37 26 +14·8 −17·0	40 08 −1·2	0 + 0·2	15·5 −6·9 51·3	100 − 9·7	
38 36 +15·1 −17·2	39 50 +14·9 −16·9	42 44 −1·1	41 + 0·2	16·0 −7·0 52·8	105 − 9·9	
41 08 +15·2 −17·1	42 31 +15·0 −16·8	45 36 −1·0	75 + 0·1	16·5 −7·1 54·3	110 −10·2	
43 59 +15·3 −17·0	45 31 +15·1 −16·7	48 47 −0·9	Mar. 23–Dec. 31	16·9 −7·2 55·8	115 •10·4	
47 10 +15·4 −16·9	48 55 +15·2 −16·6	52 18 −0·8	0 + 0·1	17·4 −7·3 57·4	120 −10·6	
50 46 +15·5 −16·8	52 44 +15·3 −16·5	56 11 −0·7	60 + 0·1	17·9 −7·4 58·9	125 −10·8	
54 49 +15·6 −16·7	57 02 +15·4 −16·4	60 28 −0·6		18·4 −7·5 60·5		
59 23 +15·7 −16·6	61 51 +15·5 −16·3	65 08 −0·5		18·8 −7·6 62·1	130 −11·1	
64 30 +15·8 −16·5	67 17 +15·6 −16·2	70 11 −0·4		19·3 −7·7 63·8	135 −11·3	
70 12 +15·9 −16·4	73 16 +15·7 −16·1	75 34 −0·3		19·8 −7·8 65·4	140 −11·5	
76 26 +16·0 −16·3	79 43 +15·8 −16·0	81 13 −0·2		20·4 −7·9 67·1	145 −11·7	
83 05 +16·1 −16·2	86 32 +15·9 −15·9	87 03 −0·1		20·9 −8·0 68·8	150 −11·9	
90 00	90 00	90 00 0·0		21·4 −8·1 70·5	155 −12·1	

App. Alt. = Apparent altitude = Sextant altitude corrected for index error and dip.
For daylight observations of Venus, see page 260.

Figure 1–3. Example of Altitude Correction Tables from Nautical Almanac

A2 ALTITUDE CORRECTION TABLES 10°-90°

OCT.—MAR. **SUN** APR.—SEPT.					
App. Alt.	Lower Limb	Upper Limb	App. Alt.	Lower Limb	Upper Limb
° ′			° ′		
9 34	+10·8	−21·5	9 39	+10·6	−21·2
9 45	+10·9	−21·4	9 51	+10·7	−21·1
9 56	+11·0	−21·3	10 03	+10·8	−21·0
10 08			10 15	+10·9	−20·0
54 49	+15·5	−16·8	57 02	+15·4	16·4
59 23	+15·6	16·7	61 51	+15·5	16·3
64 30	+15·7	16·6	67 17	+15·6	16·2
70 12	+15·8	16·5	73 16	+15·7	16·1
76 26	15·9	16·4	79 43	15·8	16·0
83 05	16·0	16·3	86 32	15·9	15·9
90 00	16·1	16·2	90 00		

DIP				
Ht. of Eye	Corrn	Ht. of Eye	Ht. of Eye	Corrn
m		ft.	m	
2·4	−2·8	8·0	1·0	−1·8
2·6		8·6	1·5	−2·2
2·8	−2·9	9·2	2·0	−2·5
3·0	−3·0	9·8	2·5	−2·8
18·4	−7·5	60·5		
18·8	−7·6	62·1	130	11·1
19·3	−7·7	63·8	135	11·3
19·8	−7·8	65·4	140	11·5
20·4	−7·9	67·1	145	11·7
20·9	−8·0	68·8	150	11·9
21·4	−8·1	70·5	155	12·1

App. Alt.   Apparent altitude

Sextant altitude corrected for index error and dip.

Figure 1–4. Portion of Altitude Correction Tables from Nautical Almanac showing Dip correction (−2.9' for Ht. of Eye of 9 ft.) and Sun's Lower Limb Refraction correction (+15.6' for App. Alt. of 69° 44.2') for observations April-September

−2.9′, the correction for all heights of eye between 8.6 and 9.2 feet. Try this yourself.

Totalling the IC and D corrections and applying them to your hs which was 69-48.5, you arrive at an intermediate figure called "Apparent Altitude" (ha), used to enter the final altitude correction table. Thus, for your sight, the ha would be 69-44.2.

For the last altitude correction, Refraction (R), you enter the SUN column of the tables (Figure 1-4). Strictly speaking, the almanac makers have lumped several corrections together which depend upon the apparent altitude but, for convenience, they are considered as a single correction, R, which makes life easier. Assuming you shot your sight in June, enter the APR-SEPT column under Lower Limb, and, with the ha of 69-44.2, find the correction of +15.6' between the appropriate critical values. Had you brought the upper limb of the sun to the horizon in your observation, the R correction would have been taken from the upper limb column and, as you will see in the table, would have been −16.2' for the same ha.

Every navigator worth his salt keeps a workbook, a standard procedure in the Navy and Merchant Marine. One easy way is with a simple form, which not only helps you remember the steps and establish a routine, but also makes it possible to check back over your work. Had you been doing this for the sun sight you're working, the workbook, so far, might look like this:

DATE	June 9
BODY	Sun, Lower Limb
hs	69-48.5
IC	−1.4
D	−2.9
ha	69-44.2
R	+15.6
Ho	69-59.8

You now have your observational information complete, and you set the Ho aside for a minute while we proceed to the second step, taking the time of the sight.

2. Timing the Sight

In piloting, the approximate time of establishing a line of position was good enough, since a small vessel is moving relatively slowly and the landmarks are fixed. In celestial, on the other hand, the bodies from which we get our position lines are constantly in motion with respect to the earth. It becomes important to get the exact time of the observation. In the case of your sun line, each second of time could mean an error of ¼ mile in your position, so it is desirable to time your sight as closely as possible to the *exact* second.

Historically, the lack of accurate time was the navigator's nemesis, but the advent of electronic watches and the availability of radio time signals all over the world has made the accuracy we need readily attainable.

Since the almanac data are presented in terms of the time at 0° longitude, called "Greenwich Mean Time" (GMT), I prefer to set my watch to Greenwich Time, five hours later than Eastern Standard, or four hours later than Eastern Daylight Saving Time. At sea, if possible, I try to reset my watch once each day with a radio time tick. If you don't choose to keep GMT on your watch, or don't wish to reset it with each radio check, you can simply compare it with the correct time, note the error, and apply the correction to your watch's reading when you time your sight.

I have found it convenient to hold my watch in the palm of my left hand while I'm taking a sight, and, at the moment of tangency, glance first at the watch time, record it and then note the sextant reading.

Continuing with our exercise, let's assume you had your watch (W) set to GMT and a recent time tick showed it to be right on. Say you took the sight just before noon, local time, and your watch read 15 hours, 56 minutes and 55 seconds. The next entry in your workbook would look like this:

W	15-56-55
corr	00
GMT	15-56-55

It is with this time that you are now ready to enter the almanac.

3. The Almanac

You have already encountered the *Nautical Almanac* informally when you opened the cover to use the altitude correction tables, but now is the time for a formal introduction. The almanac contains all the astronomical information the navigator needs to proceed with the conversion of his sight data to a useful line of position. In addition, there is a wealth of auxiliary and planning data which you will soon find of interest, and an excellent explanation section toward the end of the book.

There are actually two choices of almanacs available to the navigator: the *Nautical Almanac,* which is published once for each calendar year, and the *American Air Almanac* which is issued in several segments each covering a period of months. Both are prepared by the U.S. Naval Observatory and published by the Government Printing Office. Although some years ago there was an argument for the simplicity and presentation of the *Air Almanac,* the *Nautical Almanac* has, in recent years, adopted a straightforward, convenient format and, as you may have surmised, I prefer it, not only for the cheaper price and the convenience of the single volume, but also for its information specifically oriented to the marine navigator. As a consequence, all the examples in this book have been worked with the *Nautical Almanac.*

There are two pieces of information you want from the almanac at this point to continue working your sun sight: The Greenwich Hour Angle (GHA) of the sun, and its declination (Dec), both at the exact time of the observation. The GHA and Dec are the coordinates, like longitude and latitude are coordinates on earth, of a body on the so-called celestial sphere.

Now the celestial sphere doesn't exist in actuality but it is a convenient, imaginary sphere, concentric with the earth, with the earth at its center. All the celestial bodies are presumed to be projected upon it, and the earth's rotation from west to east causes the apparent westward motion of the bodies on the celestial sphere. Sound a little etheral? Well, don't worry about it, it's just a convenient way to get a feel for the relative motion of the heavenly bodies. More important is the understanding that the Greenwich Hour Angle (GHA) is the angular distance measured westward from the meridian of Greenwich—as is west longitude (see Figure 3-4)— while declination is the angular distance north or south of the celestial equator.

Returning to your sun sight, and assuming an arbitrary date of June 9th, you open the almanac to the appropriate page where you see the complete astronomical data displayed for a three day period. The reduced reproduction (Figure 3-1) is of the right hand page, or

G.M.T.	SUN		MOON				Lat.	Twilight		Sunrise	Moonrise			
	G.H.A.	Dec.	G.H.A.	v	Dec.	d	H.P.		Naut.	Civil		9	10	11	12

G.M.T.	G.H.A. ° '	Dec. ° '	G.H.A. ° '	v '	Dec. ° '	d '	H.P. '	Lat. °	Naut. h m	Civil h m	Sunrise h m	9 h m	10 h m	11 h m	12 h m
9 00	180 15.0	N22 52.8	142 58.0	12.9	N16 50.0	3.9	54.0	N 72	□	□	□	04 15	05 58	07 37	09 15
01	195 14.9	53.1	157 29.9	12.9	16 46.1	4.0	54.0	N 70	□	□	□	05 02	06 28	07 57	09 27
02	210 14.8	53.3	172 01.8	12.9	16 42.1	4.1	54.0	68	□	□	□	05 32	06 51	08 13	09 37
03	225 14.7	·· 53.5	186 33.7	13.0	16 38.0	4.2	54.0	66	////	////	00 31	05 55	07 08	08 25	09 45
04	240 14.6	53.7	201 05.7	13.0	16 33.8	4.2	54.0	64	////	////	01 40	06 12	07 22	08 36	09 52
05	255 14.4	53.9	215 37.7	13.0	16 29.6	4.3	54.0	62	////	////	02 14	06 27	07 34	08 45	09 58
06	270 14.3	N22 54.1	230 09.7	13.0	N16 25.3	4.4	54.0	60	////	01 03	02 39	06 39	07 44	08 52	10 03
07	285 14.2	54.4	244 41.7	13.1	16 20.9	4.5	54.0	N 58	////	01 46	02 59	06 50	07 53	08 59	10 07
08	300 14.1	54.6	259 13.8	13.1	16 16.4	4.5	54.0	56	////	02 14	03 15	06 59	08 00	09 05	10 11
F 09	315 13.9	·· 54.8	273 45.9	13.1	16 11.9	4.6	54.0	54	00 58	02 35	03 29	07 07	08 07	09 10	10 15
R 10	330 13.8	55.0	288 18.0	13.1	16 07.3	4.7	54.0	52	01 38	02 53	03 41	07 14	08 13	09 15	10 18
I 11	345 13.7	55.2	302 50.1	13.1	16 02.6	4.7	54.0	50	02 04	03 07	03 51	07 21	08 19	09 19	10 21
D 12	0 13.6	N22 55.4	317 22.2	13.2	N15 57.9	4.8	54.0	45	02 47	03 36	04 13	07 35	08 31	09 28	10 27
A 13	15 13.5	55.6	331 54.4	13.2	15 53.1	4.9	54.0	N 40	03 17	03 58	04 31	07 46	08 41	09 36	10 33
Y 14	30 13.3	55.8	346 26.6	13.2	15 48.2	5.0	54.0	35	03 40	04 16	04 45	07 56	08 49	09 43	10 37
15	45 13.2	·· 56.0	0 58.8	13.2	15 43.2	5.0	54.0	30	03 58	04 31	04 58	08 05	08 56	09 48	10 41
16	60 13.1	56.2	15 31.0	13.3	15 38.2	5.1	54.0	20	04 26	04 55	05 20	08 20	09 09	09 58	10 48
17	75 13.0	56.5	30 03.3	13.3	15 33.1	5.2	54.0	N 10	04 48	05 15	05 38	08 33	09 20	10 07	10 54
18	90 12.8	N22 56.7	44 35.6	13.3	N15 27.9	5.2	54.1	0	05 07	05 33	05 56	08 45	09 30	10 15	11 00
19	105 12.7	56.9	59 07.9	13.3	15 22.7	5.3	54.1	S 10	05 24	05 50	06 13	08 57	09 41	10 24	11 05
20	120 12.6	57.1	73 40.2	13.3	15 17.4	5.3	54.1	20	05 40	06 07	06 31	09 10	09 52	10 32	11 11
21	135 12.5	·· 57.3	88 12.5	13.4	15 12.1	5.5	54.1	30	05 56	06 26	06 52	09 25	10 05	10 42	11 18
22	150 12.4	57.5	102 44.9	13.4	15 06.6	5.5	54.1	35	06 04	06 36	07 04	09 33	10 12	10 48	11 22
23	165 12.2	57.7	117 17.3	13.4	15 01.1	5.5	54.1	40	06 13	06 47	07 18	09 43	10 20	10 54	11 27
10 00	180 12.1	N22 57.9	131 49.7	13.4	N14 55.6	5.7	54.1	45	06 23	07 00	07 34	09 54	10 30	11 02	11 32
01	195 12.0	58.1	146 22.1	13.5	14 49.9	5.6	54.1	S 50	06 35	07 16	07 54	10 08	10 41	11 11	11 38
02	210 11.9	58.3	160 54.6	13.4	14 44.2	5.7	54.1	52	06 40	07 23	08 04	10 15	10 47	11 15	11 41
03	225 11.7	·· 58.5	175 27.0	13.5	14 38.5	5.9	54.1	54	06 45	07 31	08 15	10 22	10 53	11 20	11 44
04	240 11.6	58.7	189 59.5	13.5	14 32.6	5.9	54.1	56	06 51	07 40	08 27	10 30	10 59	11 25	11 47
05	255 11.5	58.9	204 32.0	13.6	14 26.7	5.9	54.1	58	06 58	07 50	08 41	10 39	11 07	11 30	11 51
06	270 11.4	N22 59.1	219 04.6	13.5	N14 20.8	6.0	54.1	S 60	07 05	08 01	08 58	10 49	11 15	11 37	11 55

G.M.T.								Lat.	Sunset	Twilight		Moonset			
										Civil	Naut.	9	10	11	12
07	285 11.2	59.3	233 37.1	13.6	14 14.8	6.1	54.1								
S 08	300 11.1	59.5	248 09.7	13.6	14 08.7	6.2	54.1	Lat. °	Sunset h m	Civil h m	Naut. h m	9 h m	10 h m	11 h m	12 h m
A 09	315 11.0	·· 59.7	262 42.3	13.6	14 02.5	6.2	54.2								
T 10	330 10.9	22 59.9	277 14.9	13.6	13 56.3	6.2	54.2	N 72	□	□	□	00 59	00 51	00 45	00 39
U 11	345 10.7	23 00.0	291 47.5	13.7	13 50.1	6.4	54.2	N 70	□	□	□	00 11	00 20	00 24	00 25
R 12	0 10.6	N23 00.2	306 20.2	13.7	N13 43.7	6.4	54.2	68	□	□	23 56	00 00	00 07	00 11	00 14
D 13	15 10.5	00.4	320 52.9	13.6	13 37.3	6.4	54.2	66	23 35	////	////	23 38	23 54	24 05	00 05
A 14	30 10.4	00.6	335 25.5	13.7	13 30.9	6.5	54.2	64	22 20	////	////	23 24	23 42	23 57	24 10
Y 15	45 10.2	·· 00.8	349 58.2	13.8	13 24.4	6.6	54.2	62	21 45	////	////	23 11	23 33	23 50	24 06
16	60 10.1	01.0	4 31.0	13.7	13 17.8	6.6	54.2	60	21 20	22 57	////	23 01	23 25	23 45	24 02
17	75 10.0	01.2	19 03.7	13.8	13 11.2	6.7	54.2	N 58	21 00	22 13	////	22 52	23 17	23 39	23 59
18	90 09.9	N23 01.4	33 36.5	13.7	N13 04.5	6.8	54.2	56	20 44	21 45	////	22 44	23 11	23 35	23 56
19	105 09.7	01.6	48 09.2	13.8	12 57.7	6.8	54.3	54	20 30	21 24	23 03	22 36	23 05	23 31	23 54
20	120 09.6	01.8	62 42.0	13.8	12 50.9	6.8	54.3	52	20 18	21 06	22 22	22 30	23 00	23 27	23 52
21	135 09.5	·· 01.9	77 14.8	13.9	12 44.1	7.0	54.3	50	20 08	20 52	21 56	22 24	22 55	23 23	23 50
22	150 09.4	02.1	91 47.7	13.8	12 37.1	6.9	54.3	45	19 46	20 23	21 12	22 11	22 45	23 16	23 45
23	165 09.2	02.3	106 20.5	13.9	12 30.2	7.1	54.3	N 40	19 28	20 01	20 42	22 01	22 36	23 10	23 41
11 00	180 09.1	N23 02.5	120 53.4	13.8	N12 23.1	7.1	54.3	35	19 13	19 43	20 19	21 52	22 29	23 04	23 38
01	195 09.0	02.7	135 26.2	13.9	12 16.0	7.1	54.3	30	19 00	19 28	20 01	21 44	22 23	22 59	23 35
02	210 08.9	02.9	149 59.1	13.9	12 08.9	7.2	54.3	20	18 39	19 03	19 32	21 30	22 11	22 51	23 30
03	225 08.7	·· 03.0	164 32.0	14.0	12 01.7	7.3	54.4	N 10	18 20	18 43	19 10	21 18	22 02	22 44	23 25
04	240 08.6	03.2	179 05.0	13.9	11 54.4	7.3	54.4	0	18 02	18 25	18 52	21 07	21 52	22 37	23 21
05	255 08.5	03.4	193 37.9	13.9	11 47.1	7.3	54.4	S 10	17 46	18 09	18 35	20 56	21 43	22 30	23 17
06	270 08.3	N23 03.6	208 10.8	14.0	N11 39.8	7.4	54.4	20	17 28	17 52	18 19	20 44	21 33	22 22	23 12
07	285 08.2	03.8	222 43.8	14.0	11 32.4	7.5	54.4	30	17 07	17 33	18 03	20 30	21 21	22 14	23 07
08	300 08.1	04.0	237 16.8	13.9	11 24.9	7.6	54.4	35	16 55	17 23	17 54	20 21	21 15	22 09	23 04
S 09	315 08.0	·· 04.1	251 49.7	14.0	11 17.4	7.6	54.4	40	16 42	17 11	17 45	20 12	21 07	22 03	23 00
U 10	330 07.8	04.3	266 22.7	14.0	11 09.8	7.6	54.5	45	16 26	16 58	17 35	20 01	20 58	21 56	22 56
N 11	345 07.7	04.5	280 55.7	14.1	11 02.2	7.7	54.5	S 50	16 04	16 42	17 24	19 48	20 47	21 49	22 51
D 12	0 07.6	N23 04.7	295 28.8	14.0	N10 54.5	7.7	54.5	52	15 55	16 35	17 19	19 42	20 42	21 45	22 49
A 13	15 07.5	04.8	310 01.8	14.0	10 46.8	7.8	54.5	54	15 44	16 27	17 13	19 35	20 37	21 41	22 47
Y 14	30 07.3	05.0	324 34.8	14.1	10 39.0	7.8	54.5	56	15 31	16 18	17 07	19 28	20 31	21 36	22 44
15	45 07.2	·· 05.2	339 07.9	14.0	10 31.2	7.9	54.5	58	15 17	16 08	17 01	19 19	20 24	21 31	22 41
16	60 07.1	05.4	353 40.9	14.1	10 23.3	7.9	54.6	S 60	15 00	15 57	16 53	19 09	20 16	21 26	22 37
17	75 07.0	05.5	8 14.0	14.1	10 15.4	8.0	54.6								

									SUN			MOON			
18	90 06.8	N23 05.7	22 47.1	14.0	N10 07.4	8.0	54.6								
19	105 06.7	05.9	37 20.1	14.1	9 59.4	8.1	54.6		Eqn. of Time		Mer.	Mer. Pass.		Age	Phase
20	120 06.6	06.0	51 53.2	14.1	9 51.3	8.1	54.6	Day	00ʰ	12ʰ	Pass.	Upper	Lower		
21	135 06.4	·· 06.2	66 26.3	14.1	9 43.2	8.2	54.6		m s	m s	h m	h m	h m	d	
22	150 06.3	06.4	80 59.4	14.1	9 35.0	8.2	54.7	**9**	01 00	00 55	11 59	14 56	02 33	04	
23	165 06.2	06.6	95 32.5	14.1	9 26.8	8.2	54.7	10	00 49	00 43	11 59	15 41	03 19	05	◖
								11	00 37	00 31	11 59	16 26	04 04	06	
	S.D. 15.8	d 0.2	S.D.	14.7		14.8	14.8									

Figure 3–1. Example of Daily Page for Sun and Moon from Nautical Almanac

"sun side" of the 1978 *Nautical Almanac* for your date. Figure 3-2 is a full-sized excerpt of that portion of the table applicable to your sight.

If you look down the SUN column to the hour of GMT at which you took the sight (remember the GMT was 15-56-55), you will see alongside 15 hours a GHA of 45-13.2 and a Dec of 22-56.0 N. You still need to correct

Figure 3–2. Excerpt from Nautical Almanac showing Astronomical Data for the Sun on June 9 at 15 ^{h} G.M.T.

1978 JUNE 9, 10, 11 (FRI., SAT., SUN.)

G.M.T.	SUN		MOON				
	G.H.A.	Dec.	G.H.A.	v	Dec.	d	H
d h	° ′	° ′	° ′	′	° ′	′	
➤9 00	180 15.0	N22 52.8	142 58.0	12.9	N16 50.0	3.9	5
01	195 14.9	53.1	157 29.9	12.9	16 46.1	4.0	5·
02	210 14.8	53.3	172 01.8	12.9	16 42.1	4.1	54
03	225 14.7 ··	53.5	186 33.7	13.0	16 38.0	4.2	54.
04	240 14.6	53.7	201 05.7	13.0	16 33.8	4.2	54.
05	255 14.4	53.9	215 37.7	13.0	16 29.6	4.3	54.(
06	270 14.3	N22 54.1	230 09.7	13.0	N16 25.3	4.4	54.(
07	285 14.2	54.4	244 41.7	13.1	16 20.9	4.5	54.
08	300 14.1	54.6	259 13.8	13.1	16 16.4	4.5	54
F 09	315 13.9 ··	54.8	273 45.9	13.1	16 11.9	4.6	54
R 10	330 13.8	55.0	288 18.0	13.1˙	16 07.3	4.7	54
I 11	345 13.7	55.2	302 50.1	13.1	16 02.6	4.7	54
D 12	0 13.6	N22 55.4	317 22.2	13.2	N15 57.9	4.8	54
A 13	15 13.5	55.6	331 54.4	13.2	15 53.1	4.9	54
Y 14	30 13.3	55.8	346 26.6	13.2	15 48.2	5.0	54.
✦15	45 13.2 ··	56.0	0 58.8	13.2	15 43.2	5.0	54.(
16	60 13.1	56.2	15 31.0	13.3	15 38.2	5.1	54.0
17	75 13.0	56.5	30 03.3	13.3	15 33.1	5.2	54.0
18	90 12.8	N22 56.7	44 35.6	13.3	N15 27.9	5.2	54.1
19	105 12.7	56.9	59 07.9	13.3	15 22.7	5.3	54.1
			73 40.2	13.3			54.1

these figures for the 56 minutes and 55 seconds remaining. In the case of the GHA, you turn to the yellow "Increments and Corrections" pages in the back of the almanac, an excerpt of which is shown in Figure 3-3. Looking under 56 minutes, you go down the SUN column to 55 seconds where you extract the sun's increment of 14° 13.8′.

Figure 3–3. Excerpt from Nautical Almanac Increments and Corrections Tables showing Sun's increment for 56ᵐ 55ˢ

56ᵐ INCREMENTS AND CORRECTIONS

56ᵐ	SUN PLANETS	ARIES	MOON	v or Corrⁿ d		v or Corrⁿ d		v or Corrⁿ d	
s	° ′	° ′	° ′	′	′	′	′	′	′
00	14 00·0	14 02·3	13 21·7	0·0	0·0	6·0	5·7	12·0	11·3
01	14 00·3	14 02·6	13 22·0	0·1	0·1	6·1	5·7	12·1	11·4
02	14 00·5	14 02·8	13 22·2	0·2	0·2	6·2	5·8	12·2	11·5
03	14 00·8	14 03·1	13 22·4	0·3	0·3	6·3	5·9	12·3	11·6
04	14 01·0	14 03·3	13 22·7	0·4	0·4	6·4	6·0	12·4	11·7
51	14 12·8	14 15·1	13 33·9	5·1	4·8	11·1	10·5	17·1	16·1
52	14 13·0	14 15·3	13 34·1	5·2	4·9	11·2	10·5	17·2	16·2
53	14 13·3	14 15·6	13 34·4	5·3	5·0	11·3	10·6	17·3	16·3
54	14 13·5	14 15·8	13 34·6	5·4	5·1	11·4	10·7	17·4	16·4
55	14 13·8	14 16·1	13 34·9	5·5	5·2	11·5	10·8	17·5	16·5
56	14 14·0	14 16·3	13 35·1	5·6	5·3	11·6	10·9	17·6	16·6
57	14 14·3	14 16·6	13 35·3	5·7	5·4	11·7	11·0	17·7	16·7
58	14 14·5	14 16·8	13 35·6	5·8	5·5	11·8	11·1	17·8	16·8
59	14 14·8	14 17·1	13 35·8	5·9	5·6	11·9	11·2	17·9	16·9
60	14 15·0	14 17·3	13 36·1	6·0	5·7	12·0	11·3	18·0	17·0

The correction to declination is even easier. As you can see by inspection (Figure 3-2), on this date the declination is increasing, but only at the rate of 0.2′ per hour. Since you are practically at 16 hours, a +0.2′ correction, and a resulting Dec of 22-56.2 N would apply. Note that while the declination is *increasing* on this particular date, it doesn't always, so it's important that you note the sign and apply it correctly when figuring declination.

You should enter these calculations in your workbook, which would look as follows:

GMT	15-56-55
gha(15ʰ)	45-13.2
incr (56ᵐ55ˢ)	14-13.8
GHA	59-27.0
Dec	22-56.2 N

Since you now want the sun's Local Hour Angle (LHA)—the same kind of measurement as GHA except that it starts from your own meridian—to enter the sight reduction table in the next step, we must apply the observer's longitude to the GHA we have calculated. For convenience select a longitude near your dead reckoning longitude so that the LHA will work out to a whole degree. Such a selected longitude is called your "Assumed Longitude" and is abbreviated, aλ. Since the sun's apparent motion is westward, you must subtract your Assumed Longitude from GHA if the longitude is west, and add it if east. The diagram in Figure 3-4 should help to clarify the relationship between GHA, LHA, and the observer's Assumed Longitude.

Perhaps it's even easier to remember the simple formula:

$$\text{LHA} = \text{GHA} \begin{array}{c} -\text{West} \\ +\text{East} \end{array} \text{Longitude}$$

To continue, say the longitude of your DR position at the time of the sight was 70° 14' West. Since the GHA was 59-27.0, a convenient Assumed Longitude, to make the

Figure 3–4. Greenwich Hour Angle is measured westward through 360° from the Greenwich Meridian to that of the Celestial Body; Longitude from the Greenwich Meridian east or west through 180° to that of the Observer; Local Hour Angle through 360° from the Observer's Meridian west to that of the Celestial Body

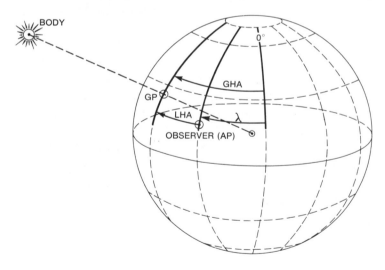

LHA come out to a whole degree, would be 70-27.0 W. But, as we must *subtract* west longitude and, as it is greater than the GHA, what do we do? Checkmate? Not at all. This is just an example of the occasional case in which in which west longitude exceeds GHA and you simply add 360° to your GHA and proceed from there. In a like manner, if you find yourself in east longitudes and the sum of your GHA and aλ *exceeds* 360°, the correct LHA can be found by *subtracting* 360° from the total.

The continuing entries in your workbook now look like this:

GHA   59-27.0
         +360-00
          419-27.0
aλ(W) − 70-27.0
LHA   349

One further detail which must be attended to before moving to the next step—calculating the computed altitude—is to establish the remaining figure that the sight reduction table requires, besides LHA and Dec: Latitude. Just assume the whole degree of latitude nearest your DR position. If you were in approximately 40° 43' North at the time of your sight, your Assumed Latitude (aL) would be 41° N.

You now have both coordinates of your Assumed Position (AP) which will be of use again in the last step: plotting the line of position. In your example, those we have assumed are:

aL   41 N
aλ   70-27 W

and the three entries needed to enter the sight reduction
table are displayed in your workbook as:

LHA	349
Dec	22-56.2 N
aL	41 N

4. Calculating the Computed Altitude

Step four is to calculate what the altitude and azimuth (the true bearing) of the body you observed *should* be from your Assumed Position. Then you will be able to compare this Computed Altitude (Hc) with the Observed Altitude (Ho) which you have determined earlier, to give you the information needed for your plot.

You can obtain the Hc from the three figures for LHA, Dec, and aL by the formulae of spherical trigonometry, but if the name doesn't scare you, the work required will. The computation was so burdensome until the coming of electronic calculators that for two centuries navigators have concentrated on providing easier methods of solution. Consequently, "inspection tables" were born and virtually all navigators today utilize them.

Over the years, a vast variety of these "Sight Reduction Tables" have appeared and, like fishermen with their baits, most navigators have their favorites. Enough still like the old "Dreisonstok" (Pub. No. 208) which appeared first 50 years ago, that it still finds its way into print from time to time, while others prefer "Ageton" (Pub. No. 211) because it is a very compact table, handy to carry about. Many merchant mariners still use Pub. No. 214, which was standard during World War II, and was the first of the truly modern, inspection tables. This has now been replaced by Pub. No. 229, *Sight Reduction Tables for Marine Navigation,* which is currently standard in the Navy. Pub. No. 249, *Sight Reduction Tables for Air Navigation* is also being published at present. Both tables are sold by the Government Printing Office, the Defense Mapping Agency Hydrographic Center (DMAHC) and their agents.

Which is for you? You may take your choice between the completeness and accuracy of Pub. No. 229, or the speed and simplicity of Pub. No. 249. I recommend the latter as it is a little easier to grasp, cheaper to buy, quicker to use and its accuracy is consistent with the accuracy you can expect with observations made from a small boat. In any event, the techniques of using the two tables are generally similar in concept, so that understanding of one will make it easier for you to learn the other.

Don't be fooled by the "Air Navigation" in the title of Pub. No. 249, as it is completely compatible with the *Nautical Almanac,* and is finding increasing favor with marine navigators. It comes in three volumes: Vol. I is for selected stars, and is discussed in Chapter 9; Vol. II is for the remainder of the bodies in the solar system for use in Latitudes 0°-39°; and Vol. III is similar to II, but for Latitudes 40°-89°. Vol. I is updated every five years, while Vols. II and III remain the same and are simply reprinted from time to time.

For reducing your sun sight, you will need Vol. III (Latitudes 40°-89°) which you will enter with the three figures for Latitude, Declination, and LHA, last noted in your workbook. Open the table to the pages for 41°, your Assumed Latitude, and select the one headed "DECLINATION (15°-29°) *SAME* NAME AS LATITUDE" since, you will remember, your Dec was 22-56.2 *North,* the same "name" as your Assumed Latitude of 41° N. A reproduction of this page, in greatly reduced size, is shown as Figure 4-1.

To follow your exercise in the table, the excerpt from the aforementioned page, shown in Figure 4-2, will provide the figures you need.

Going down the LHA column, in this case on the right hand side of the page, to your LHA of 349, you will find under Declination 22° an Hc (Computed Altitude) of 68-51. This is called the "Tabulated Altitude", abbreviated Tab Hc, and is entered in your workbook along with the figures for *d* and Z which we'll come to in a minute. Note that *d* in this context represents the difference between the accompanying Hc and that for the next larger degree of declination. You do have to take into account the remaining 56.2 minutes of declination, which can be done by multiplying the tabular differ-

Figure 4–1. Example of table from Pub. No. 249, Sight Reduction Tables for Air Navigation

DECLINATION (15°–29°) SAME NAME AS LATITUDE

N. Lat. {LHA greater than 180°....... Zn=Z
{LHA less than 180°....... Zn=360−Z

LHA	15° Hc d Z	16° Hc d Z	17° Hc d Z	18° Hc d Z	19° Hc d Z	20° Hc d Z	21° Hc d Z	22° Hc d Z	23° Hc d Z	24° Hc d Z	25° Hc d Z	26° Hc d Z	27° Hc d Z	28° Hc d Z	29° Hc d Z	LHA
0	64 00 +60 180	65 00 +60 180	66 00 +60 180	67 00 +60 180	68 00 +60 180	69 00 +60 180	70 00 +60 180	71 00 +60 180	72 00 +60 180	73 00 +60 180	74 00 +60 180	75 00 +60 180	76 00 +60 180	77 00 +60 180	78 00 +60 180	360
1	63 59 60 178	64 59 60 178	65 59 60 178	66 59 60 178	67 59 60 178	68 59 60 178	69 59 60 177	70 59 60 177	71 59 60 177	72 59 60 177	73 59 60 177	74 59 60 177	75 59 59 177	76 58 60 176	77 58 59 176	359
2	63 57 60 176	64 57 60 176	65 56 60 176	66 56 60 176	67 56 60 175	68 56 60 175	69 56 60 175	70 56 59 175	71 55 60 174	72 55 59 174	73 54 60 174	74 54 59 173	75 53 59 173	76 52 59 172	77 51 59 172	358
3	63 52 59 174	64 52 59 174	65 51 60 174	66 51 59 173	67 51 59 173	68 50 59 173	69 50 59 172	70 49 59 172	71 48 59 171	72 47 59 171	73 46 59 170	74 45 59 170	75 44 58 169	76 42 59 169	77 41 58 168	357
4	63 46 60 171	64 46 59 171	65 45 59 171	66 44 59 170	67 44 58 170	68 43 58 169	69 42 59 169	70 41 58 168	71 39 58 168	72 38 58 167	73 36 58 166	74 34 58 166	75 32 58 165	76 30 57 164	77 27 58 164	356
5	63 38 +60 169	64 38 +59 169	65 37 +59 168	66 36 +59 168	67 35 +58 167	68 34 +59 167	69 33 +58 166	70 32 +58 166	71 31 +57 165	72 30 +58 164	73 28 +57 164	74 26 +57 163	75 24 +56 162	76 22 +56 161	77 20 +55 160	355
6	63 29 59 167	64 28 59 167	65 27 59 166	66 26 59 166	67 25 58 165	68 23 59 164	69 22 58 164	70 20 58 163	71 18 58 162	72 17 57 162	73 14 58 161	74 12 57 160	75 09 57 159	76 06 57 158	77 03 57 157	354
7	63 18 60 165	64 17 60 164	65 16 59 164	66 15 59 163	67 13 59 162	68 12 58 162	69 10 58 161	70 08 58 160	71 06 57 159	72 03 58 159	73 01 57 158	73 58 57 157	74 55 57 156	75 52 56 155	76 48 56 154	353
8	63 06 59 162	64 05 59 162	65 04 59 161	66 03 59 161	67 01 58 160	68 00 58 159	68 58 58 158	69 56 57 157	70 53 57 156	71 50 57 156	72 47 57 155	73 44 56 154	74 40 56 153	75 36 56 152	76 32 55 151	352
9	62 51 59 161	63 50 59 160	64 49 59 159	65 48 59 159	66 46 58 158	67 45 58 157	68 43 58 156	69 41 57 155	70 38 57 154	71 35 56 153	72 31 57 152	73 28 56 152	74 24 55 150	75 19 56 149	76 15 55 148	351
10	62 35 +58 158	63 35 +58 158	64 34 +58 157	65 33 +58 156	66 31 +58 155	67 30 +57 154	68 28 +57 154	69 25 +57 153	70 22 +56 152	71 18 +57 151	72 15 +55 150	73 10 +56 149	74 06 +55 148	75 01 +55 147	75 56 +54 146	350
11	62 18 59 156	63 17 58 155	64 15 59 155	65 14 58 154	66 12 58 153	67 10 57 152	68 07 58 151	69 05 57 150	70 02 56 149	70 58 56 148	71 54 56 147	72 50 55 146	73 45 55 145	74 40 54 144	75 34 54 143	349
12	61 59 58 154	62 58 58 153	63 56 58 152	64 55 58 152	65 53 57 151	66 50 57 150	67 48 57 149	68 45 56 148	69 41 57 147	70 38 55 146	71 33 56 145	72 29 55 144	73 24 54 142	74 18 54 141	75 12 53 140	348
13	61 39 58 151	62 37 58 151	63 35 58 150	64 33 58 149	65 31 57 148	66 28 57 147	67 25 57 146	68 22 56 145	69 18 56 144	70 14 56 143	71 10 55 142	72 05 54 140	72 59 54 139	73 53 53 138	74 46 53 137	347
14	61 18 58 149	62 15 58 148	63 13 58 147	64 11 57 146	65 08 57 145	66 05 57 144	67 02 56 143	67 58 56 142	68 54 56 141	69 50 55 140	70 45 55 139	71 40 54 138	72 34 53 137	73 27 53 135	74 20 52 134	346
15	60 55 +57 147	61 53 +57 146	62 50 +57 145	63 47 +57 144	64 44 +57 143	65 41 +56 142	66 37 +56 141	67 33 +56 140	68 29 +55 139	69 24 +55 138	70 19 +54 137	71 13 +54 135	72 07 +53 134	72 59 +53 133	73 52 +52 132	345
16	60 30 57 145	61 27 57 144	62 24 57 143	63 21 57 142	64 18 56 141	65 14 56 140	66 10 56 139	67 06 55 138	68 01 55 137	68 56 55 136	69 51 54 134	70 45 53 133	71 38 53 132	72 31 52 131	73 23 51 130	344
17	60 06 56 143	61 03 57 142	62 00 56 141	62 56 57 140	63 53 56 139	64 49 56 138	65 45 55 137	66 40 55 136	67 35 55 135	68 30 54 133	69 24 53 132	70 17 53 131	71 10 52 130	72 02 52 128	72 54 51 127	343
18	59 40 57 141	60 37 56 140	61 33 56 139	62 30 56 138	63 26 55 137	64 21 56 136	65 17 55 135	66 12 54 134	67 06 54 132	68 00 54 131	68 54 53 130	69 47 52 129	70 39 52 127	71 31 51 126	72 22 51 125	342
19	59 13 56 139	60 09 56 138	61 05 56 137	62 01 56 136	62 57 55 135	63 52 55 134	64 47 54 132	65 41 55 131	66 36 53 130	67 29 54 129	68 23 52 128	69 15 52 126	70 07 51 125	70 58 51 124	71 49 50 122	341
20	58 45 +56 137	59 41 +56 136	60 37 +55 135	61 32 +56 134	62 28 +55 133	63 23 +54 132	64 17 +54 131	65 11 +54 129	66 05 +53 128	66 58 +53 127	67 51 +52 126	68 43 +52 124	69 35 +50 123	70 25 +51 122	71 16 +49 120	340
21	58 16 56 135	59 12 55 134	60 07 55 133	61 02 55 132	61 57 54 131	62 51 54 130	63 45 54 128	64 39 53 127	65 32 53 126	66 25 52 125	67 17 52 123	68 09 51 122	69 00 51 120	69 51 49 119	70 40 50 118	339
22	57 47 55 133	58 42 55 132	59 37 55 131	60 32 54 130	61 26 54 129	62 20 54 128	63 14 53 126	64 07 53 125	65 00 52 124	65 52 52 122	66 44 51 121	67 35 51 120	68 26 50 118	69 16 49 117	70 05 49 115	338
23	57 17 55 132	58 12 54 130	59 06 55 129	60 01 54 128	60 55 53 127	61 48 54 126	62 42 53 124	63 35 52 123	64 27 52 122	65 19 51 120	66 10 51 119	67 01 50 117	67 51 50 116	68 41 49 114	69 30 48 112	337
24	56 47 55 130	57 42 54 129	58 36 54 128	59 30 54 126	60 24 53 125	61 17 53 124	62 10 52 122	63 02 52 121	63 54 51 120	64 45 51 118	65 36 50 117	66 26 50 115	67 16 48 114	68 04 49 112	68 53 47 110	336
25	56 09 +49 127	56 58 +54 127	57 52 +54 126	58 46 +53 125	59 40 +53 123	60 33 +52 122	61 25 +52 121	62 17 +52 119	63 09 +51 118	63 59 +50 117	64 49 +50 115	65 39 +48 114	66 27 +49 112	67 16 +47 110	68 03 +47 109	335
26	55 35 54 126	56 29 53 124	57 22 53 123	58 16 53 122	59 09 52 121	60 01 52 119	60 53 52 118	61 45 51 117	62 36 50 115	63 26 50 114	64 16 49 112	65 05 49 111	65 54 47 109	66 41 48 108	67 29 46 106	334
27	55 01 53 124	55 54 53 123	56 47 53 122	57 40 52 121	58 32 52 119	59 24 52 118	60 16 51 117	61 07 50 115	61 57 50 114	62 47 49 112	63 36 49 111	64 25 47 109	65 12 48 108	66 00 46 106	66 46 46 105	333
28	54 26 53 122	55 19 53 121	56 12 52 120	57 04 52 119	57 56 51 117	58 47 51 116	59 38 51 115	60 29 49 113	61 18 50 112	62 08 48 110	62 56 48 109	63 44 47 107	64 31 46 106	65 17 46 104	66 03 45 103	332
29	53 51 53 121	54 44 52 119	55 36 52 118	56 28 51 117	57 19 51 116	58 10 50 114	59 00 50 113	59 50 49 112	60 39 49 110	61 28 48 109	62 16 47 107	63 03 47 106	63 50 45 104	64 35 45 103	65 20 44 101	331
30	53 14 +52 119	54 06 +52 118	54 58 +51 117	55 49 +51 115	56 40 +51 114	57 31 +50 113	58 21 +49 111	59 10 +49 110	59 59 +48 109	60 47 +48 107	61 35 +46 106	62 21 +46 104	63 07 +46 103	63 53 +44 101	64 37 +44 100	330
31	52 37 52 118	53 29 51 116	54 20 51 115	55 11 50 114	56 01 50 113	56 51 50 111	57 41 49 110	58 30 48 108	59 18 48 107	60 06 47 106	60 53 46 104	61 39 45 103	62 24 45 101	63 09 44 100	63 53 43 98	329
32	52 00 51 116	52 51 51 115	53 42 51 113	54 33 49 112	55 22 50 111	56 12 49 110	57 01 48 108	57 49 48 107	58 37 47 106	59 24 46 104	60 10 46 103	60 56 45 101	61 41 43 100	62 24 44 98	63 08 42 97	328
33	51 22 51 115	52 13 51 114	53 04 50 112	53 54 50 111	54 44 49 110	55 33 48 108	56 21 48 107	57 09 47 106	57 56 47 104	58 43 46 103	59 29 45 101	60 14 44 100	60 58 44 98	61 42 43 97	62 25 42 96	327
34	50 44 +51 113	51 35 +50 112	52 25 +50 111	53 15 +49 110	54 04 +49 108	54 53 +48 107	55 41 +47 106	56 28 +47 104	57 15 +46 103	58 01 +45 102	58 46 +45 100	59 31 +43 99	60 14 +43 97	60 57 +42 96	61 39 +41 95	326
35	50 05 50 112	50 55 50 111	51 45 49 109	52 34 49 108	53 23 48 107	54 11 48 106	54 59 47 104	55 46 46 103	56 32 46 102	57 18 45 100	58 03 44 99	58 47 44 97	59 31 42 96	60 13 42 95	60 55 41 93	325
36	49 26 49 110	50 15 49 109	51 04 49 108	51 53 48 107	52 41 48 105	53 29 47 104	54 16 46 103	55 02 46 102	55 48 45 100	56 33 45 99	57 18 43 97	58 01 43 96	58 44 42 95	59 26 41 93	60 07 40 92	324
37	48 46 49 109	49 35 49 108	50 24 48 106	51 12 48 105	52 00 47 104	52 47 47 103	53 34 46 102	54 20 45 100	55 05 44 99	55 49 44 98	56 33 43 96	57 16 42 95	57 58 42 93	58 40 40 92	59 20 40 91	323
38	48 06 49 108	48 55 48 107	49 43 48 105	50 31 47 104	51 18 47 103	52 05 46 102	52 51 46 100	53 37 44 99	54 21 44 98	55 05 43 97	55 48 43 95	56 30 41 94	57 11 41 92	57 52 40 91	58 32 39 90	322
39	47 25 48 106	48 13 48 105	49 01 47 104	49 48 47 103	50 35 47 102	51 22 45 100	52 07 45 99	52 52 45 98	53 37 43 97	54 20 43 95	55 03 42 94	55 45 41 93	56 26 40 91	57 06 40 90	57 46 38 89	321
40	46 44 +48 105	47 32 +47 104	48 19 +47 103	49 06 +47 102	49 53 +46 101	50 39 +45 99	51 24 +45 98	52 09 +44 97	52 53 +43 96	53 36 +43 94	54 19 +41 93	55 00 +41 92	55 41 +40 90	56 21 +39 89	57 00 +38 88	320
41	46 03 47 104	46 50 47 103	47 37 46 102	48 23 46 101	49 09 46 99	49 55 45 98	50 40 44 97	51 24 44 96	52 08 43 95	52 51 42 93	53 33 42 92	54 15 40 91	54 55 40 89	55 35 39 88	56 14 38 87	319
42	45 22 47 103	46 09 46 102	46 55 46 101	47 41 45 100	48 26 45 98	49 11 45 97	49 56 44 96	50 40 43 95	51 23 42 94	52 05 42 92	52 47 41 91	53 28 40 90	54 08 39 89	54 47 39 87	55 26 37 86	318
43	44 41 46 102	45 27 46 101	46 13 46 100	46 59 45 98	47 44 44 97	48 28 44 96	49 12 44 95	49 56 42 94	50 38 42 92	51 20 41 91	52 01 40 90	52 41 40 89	53 21 39 87	54 00 38 86	54 38 37 85	317
44	43 59 46 100	44 45 46 99	45 31 45 98	46 16 45 97	47 01 44 96	47 45 43 95	48 28 43 94	49 11 42 92	49 53 41 91	50 34 41 90	51 15 40 89	51 55 39 88	52 34 38 86	53 12 37 85	53 49 37 84	316
45	43 15 +46 99	44 01 +45 98	44 46 +45 97	45 31 +44 96	46 15 +44 95	46 59 +43 94	47 42 +42 93	48 24 +42 91	49 06 +41 90	49 47 +40 89	50 27 +40 88	51 07 +38 86	51 45 +38 85	52 23 +37 84	53 00 +36 83	315
46	42 33 45 98	43 18 45 97	44 03 45 96	44 48 44 95	45 32 43 94	46 15 43 93	46 58 42 91	47 40 41 90	48 21 41 89	49 02 40 88	49 42 39 87	50 21 39 85	50 59 38 84	51 37 36 83	52 13 36 82	314
47	41 50 45 97	42 35 45 96	43 20 44 95	44 04 43 94	44 47 43 93	45 30 43 92	46 13 41 90	46 54 41 89	47 35 40 88	48 15 40 87	48 55 38 86	49 33 38 84	50 11 37 83	50 48 36 82	51 24 35 81	313
48	41 07 45 96	41 52 44 95	42 36 44 94	43 20 43 93	44 03 43 92	44 46 42 91	45 28 41 89	46 09 40 88	46 49 40 87	47 29 39 86	48 08 38 85	48 46 37 83	49 23 37 82	50 00 35 81	50 35 35 80	312
49	40 23 44 95	41 07 44 94	41 51 44 93	42 35 42 92	43 17 42 91	44 00 41 89	44 41 41 88	45 22 40 87	46 02 39 86	46 41 39 85	47 20 38 84	47 58 37 82	48 35 36 81	49 11 35 80	49 46 35 79	311
50	39 40 +44 94	40 24 +43 93	41 07 +43 92	41 50 +43 91	42 33 +42 90	43 15 +41 88	43 56 +40 87	44 36 +40 86	45 16 +39 85	45 55 +38 84	46 33 +38 83	47 11 +36 81	47 47 +36 80	48 23 +35 79	48 58 +34 78	310
51	38 57 43 92	39 40 43 91	40 23 43 90	41 06 42 89	41 48 41 88	42 29 41 87	43 10 40 86	43 50 39 85	44 29 39 84	45 08 38 83	45 46 37 81	46 23 36 80	46 59 36 79	47 35 34 78	48 09 34 77	309
52	38 13 43 91	38 56 43 90	39 39 42 89	40 21 42 88	41 03 41 87	41 44 40 86	42 24 40 85	43 04 39 84	43 43 38 83	44 21 38 82	44 59 37 80	45 36 35 79	46 11 35 78	46 46 34 77	47 20 33 76	308
53	37 29 43 90	38 12 42 89	38 54 42 88	39 36 41 87	40 17 41 86	40 58 40 85	41 38 39 84	42 17 39 83	42 56 38 82	43 34 37 81	44 11 37 80	44 48 35 78	45 23 35 77	45 58 34 76	46 32 33 75	307
54	36 44 42 89	37 26 42 88	38 08 42 87	38 50 41 86	39 31 41 85	40 12 39 84	40 51 39 83	41 30 38 82	42 08 38 81	42 46 37 80	43 23 36 79	43 59 35 77	44 34 34 76	45 08 34 75	45 42 32 74	306
55	35 59 +42 88	36 41 +42 87	37 23 +41 86	38 04 +41 85	38 45 +40 84	39 25 +40 83	40 05 +39 82	40 44 +37 81	41 21 +38 80	41 59 +36 79	42 35 +36 78	43 11 +34 76	43 45 +34 75	44 19 +33 74	44 52 +33 73	305
56	35 14 42 87	35 56 41 86	36 37 41 85	37 18 41 84	37 59 40 83	38 39 39 82	39 18 39 81	39 57 37 80	40 34 37 79	41 11 37 78	41 48 35 77	42 23 35 75	42 58 33 74	43 31 33 73	44 04 32 72	304
57	34 28 42 86	35 10 41 85	35 51 41 84	36 32 40 83	37 12 40 82	37 52 39 81	38 31 38 80	39 09 38 79	39 47 37 78	40 24 36 77	41 00 35 76	41 35 34 74	42 09 33 73	42 42 33 72	43 15 31 71	303
58	33 42 42 85	34 24 41 84	35 05 41 83	35 46 40 82	36 26 40 81	37 06 38 80	37 44 38 79	38 22 37 78	38 59 36 77	39 35 36 76	40 11 35 75	40 46 33 73	41 19 33 72	41 52 32 71	42 24 31 70	302
59	32 56 41 84	33 37 41 83	34 18 41 82	34 59 40 81	35 39 39 80	36 18 39 79	36 57 37 78	37 34 37 77	38 11 36 76	38 47 35 75	39 22 35 74	39 57 33 72	40 30 32 71	41 02 32 70	41 34 31 69	301
60	32 10 +41 83	32 51 +40 82	33 31 +41 81	34 12 +39 80	34 51 +40 79	35 31 +38 78	36 09 +38 77	36 47 +36 76	37 23 +36 75	37 59 +35 74	38 34 +35 73	39 09 +33 72	39 42 +32 71	40 14 +31 70	40 45 +31 69	300
61	31 23 41 82	32 04 40 81	32 44 40 80	33 24 40 79	34 04 39 78	34 43 38 77	35 21 37 76	35 58 37 75	36 35 35 74	37 10 35 73	37 45 34 72	38 19 33 71	38 52 32 70	39 24 31 69	39 55 30 68	299
62	30 36 41 81	31 17 40 80	31 57 40 79	32 37 39 78	33 16 39 77	33 55 37 76	34 32 37 75	35 09 36 74	35 45 35 73	36 20 35 72	36 55 34 71	37 29 32 70	38 01 32 69	38 33 31 68	39 04 30 67	298
63	29 49 41 80	30 30 40 79	31 10 39 78	31 49 39 77	32 28 39 76	33 07 37 75	33 44 37 74	34 21 35 73	34 56 35 72	35 31 34 71	36 05 34 70	36 39 32 69	37 11 31 68	37 42 31 67	38 13 30 66	297
64	29 03 40 79	29 43 40 78	30 23 39 77	31 02 39 76	31 41 38 75	32 19 37 74	32 56 36 73	33 32 36 72	34 08 34 71	34 42 34 70	35 16 33 69	35 49 32 68	36 21 31 67	36 52 30 66	37 22 29 65	296
65	28 15 +40 78	28 55 +40 77	29 35 +39 76	30 14 +38 75	30 52 +38 74	31 30 +37 73	32 07 +35 72	32 42 +36 71	33 18 +34 70	33 52 +34 69	34 26 +32 68	34 58 +32 67	35 30 +30 66	36 00 +30 65	36 30 +29 64	295
66	27 28 40 77	28 08 40 76	28 48 39 75	29 27 38 74	30 05 38 73	30 43 36 72	31 19 36 71	31 55 35 70	32 30 34 69	33 04 33 68	33 37 33 67	34 10 31 66	34 41 31 65	35 12 29 64	35 41 29 63	294
67	26 41 40 76	27 21 39 75	28 00 39 74	28 39 38 73	29 17 37 72	29 54 37 71	30 31 35 70	31 06 35 69	31 41 34 68	32 15 33 67	32 48 32 66	33 20 31 65	33 51 30 64	34 21 29 63	34 50 28 62	293
68	25 53 40 75	26 33 39 74	27 12 39 73	27 51 38 72	28 29 37 71	29 06 37 70	29 43 35 69	30 18 34 68	30 52 34 67	31 26 32 66	31 58 32 65	32 30 30 64	33 00 30 63	33 30 29 62	33 59 28 61	292
69	25 06 39 74	25 45 40 73	26 25 38 72	27 03 38 71	27 41 37 70	28 18 36 69	28 54 36 68	29 30 34 67	30 04 33 66	30 37 33 65	31 10 31 64	31 41 30 63	32 11 30 62	32 41 28 61	33 09 28 60	291

S. Lat. {LHA greater than 180°....... Zn=180−Z
{LHA less than 180°....... Zn=180+Z

DECLINATION (15°–29°) SAME NAME AS LATITUDE

LAT 41°

DECLINATION (15°-29°) SAME NAME AS LATITUDE

LHA	20° Hc	d	Z	21° Hc	d	Z	√22° Hc	d	Z	23° Hc	d	Z	24° Hc	d	Z	25° Hc	d	Z	LHA
	° ′	′	°	° ′	′	°	° ′	′	°	° ′	′	°	° ′	′	°	° ′	′	°	
0	69 00	+60	180	70 00	+60	180	71 00	+60	180	72 00	+60	180	73 00	+60	180	74 00	+60	180	360
1	68 59	60	177	69 59	60	177	70 59	60	177	71 59	60	177	72 59	60	177	73 59	60	177	359
2	68 56	60	175	69 56	60	175	70 56	59	174	71 55	60	174	72 55	60	174	73 55	60	173	358
3	68 51	59	172	69 50	60	172	70 50	60	172	71 50	59	171	72 49	59	171	73 48	60	170	357
4	68 44	59	170	69 43	59	169	70 42	59	169	71 41	59	168	72 40	59	168	73 39	59	167	356
5	68 34	+59	167	69 33	+59	167	70 32	+59	166	71 31	+59	165	72 30	+58	165	73 28	+58	164	355
6	68 23	59	165	69 22	58	164	70 20	58	163	71 18	59	163	72 17	57	162	73 14	58	161	354
7	68 10	58	162	69 08	58	161	70 06	58	161	71 04	57	160	72 01	57	159	72 58	57	158	353
8	67 55	58	160	68 53	57	159	69 50	57	158	70 47	57	157	71 44	56	156	72 40	56	155	352
9	67 39	57	157	68 36	56	156	69 32	57	156	70 29	56	155	71 25	55	153	72 20	55	152	351
10	67 20	+57	155	68 17	+56	154	69 13	+55	153	70 08	+55	152	71 03	+55	151	71 58	+54	149	350
11	67 00	56	153	67 56	55	152	68 51	55	151	69 46	54	150	70 40	54	148	71 34	53	147	349
12	66 39	55	151	67 34	54	149	68 28	54	148	69 22	54	147	70 16	52	146	71 08	53	144	348
13	66 16	54	148	67 10	54	147	68 04	53	146	68 57	52	145	69 49	52	143	70 41	51	142	347
14	65 51	54	146	66 45	53	145	67 38	52	144	68 30	52	143	69 22	51	141	70 13	50	140	346
15	65 26	+52	144	66 18	+52	143	67 10	+52	142	68 02	+51	140	68 53	+50	139	69 43	+49	137	345
16	64 58	52	142	65 50	52	141	66 42	50	140	67 32	50	138	68 22	49	137	69 11	49	135	344
17	64 30					139	66 12							48	135	68 39	47		

Figure 4-2. Excerpt from Table of Computed Altitudes and Azimuths, Pub. No. 249, Vol III, for Latitude 41°, Declination 22°—Same Name as Latitude—and Local Hour Angle, 349

ence *d* from the table (+55) by the incremental *degrees* of declination ($^{56.2}/_{60}$) and applying the product to Tab Hc. Even easier, use Table 5—Correction to Tabulated Altitude for Minutes of Declination—which is found at the back of the sight reduction table. Figure 4-3 shows a portion of this table.

To use Table 5, enter the heading with the value of *d* (55) and descend the vertical column to the entry op-

posite the minutes of declination (56). That figure (51')
is the correction to be applied to your Tab Hc. Notice
that sometimes the value of d may be negative, and it
will be so indicated in the table. Always be careful to
apply the correction to Tab Hc with the proper sign.

*Figure 4–3. Excerpt from Table 5, Pub. No. 249, showing
Correction to Tabulated Altitude for d of 55, and Incremental Minutes, 56*

TABLE 5.—Correction to Tabulated Altitude

43 44 45	46 47 48	49 50 51	52 53 54	55 56 57	58 59 60	$\dfrac{d}{'}$
0 0 0	0 0 0	0 0 0	0 0 0	0 0 0	0 0 0	0
1 1 1	1 1 1	1 1 1	1 1 1	1 1 1	1 1 1	1
1 1 2	2 2 2	2 2 2	2 2 2	2 2 2	2 2 2	2
2 2 2	2 2 2	2 2 3	3 3 3	3 3 3	3 3 3	3
3 3 3	3 3 3	3 3 3	3 4 4	4 4 4	4 4 4	4
4 4 4	4 4 4	4 4 4	4 4 4	5 5 5	5 5 5	5
4 4 4	5 5 5	5 5 5	5 5 5	6 6 6	6 6 6	6
5 5 5	5 5 6	6 6 6	6 6 6	6 7 7	7 7 7	7
6 6 6	6 6 6	7 7 7	7 7 7	7 7 8	8 8 8	8
34 35 36	37 38 38	39 40 . .	. . 43	44 45 46	46 47 . .	. .
35 36 37	38 38 39	40 41 42	42 43 44	45 46 47	47 48 49	49
36 37 38	38 39 40	41 42 42	43 44 45	46 47 48	48 49 50	50
37 37 38	39 40 41	42 42 43	44 45 46	47 48 48	49 50 51	51
37 38 39	40 41 42	42 43 44	45 46 47	47 48 48	49 51 52	52
38 39 40	41 42 42	43 44 45	46 47 48	49 49 50	51 52 53	53
39 40 40	41 42 43	44 45 46	47 48 49	50 50 51	52 53 54	54
39 40 41	42 43 44	45 46 47	48 49 50	50 51 52	53 54 55	55
40 41 42	43 44 45	46 47 48	49 49 50	51 52 53	54 55 56	56
41 42 43	44 45 46	47 48 48	49 50 51	52 53 54	55 56 57	57
42 43 44	44 45 46	47 48 49	50 51 52	53 54 55	56 57 58	58
42 43 44	45 46 47	48 49 50	51 52 53	54 55 56	57 58 59	59

Your workbook should now contain the following additional entries:

Tab Hc	68-51
corr	+51
Hc	69-42

You will recall that earlier you noted Z from the table which was 151°. Now, following the rules in the corner of the page, you see that in North Latitudes, when the LHA is greater than 180°, as yours was, Zn which is the abbreviation for Azimuth or *true* bearing, is equal to Z. Thus, for your sight, your Zn, which we will carry to the next step, is 151° True.

5. Altitude Difference and Azimuth

Step five is easy. Compare the Hc you have just finished calculating with the Ho which you obtained in Chapter 1; the difference, in the case of your sun observation, being 17.8′. This is called the "Altitude Difference" or "Intercept", and is abbreviated "a". For purposes of plotting you need to know whether the intercept is to be plotted "toward"—that is, in the direction of the azimuth—or "away," the reciprocal bearing. The rule states: "Ho greater, Toward; Hc greater, Away." Since, in your exercise, Ho is greater, the plot will be *toward* the azimuth direction you have established as 151° True. Your workbook should show the following as its final entries:

Hc	69-42.0
Ho	69-59.8
a	17.8′ Toward
Zn	151°

You have now completed your sun sight, and all the tabular work in reducing it. For purposes of summary and review, let me restate the problem in "book" terms and show you a completed workform which you can use to retrace your steps.

The Problem:
    A navigator in DR position Lat 40° 43′ North, Long 70° 14′ West, on June 9, 1978, makes an observation of the sun's lower limb at 15-56-55 Greenwich Mean Time. The sextant altitude is 69-48.5, and the sextant has an Index Correction of −1.4. The observer's height of eye is nine feet.

Required:
    The intercept, whether "toward" or "away", and the true azimuth.

Answer:
    Intercept (a): 17.8′ Toward; True Azimuth 151°
    When you have mastered this, and the procedure has become routine to you, you have achieved the first rank of Celestial Navigator and are entitled to welcome others to the World of Celestial Navigation, Yachtsman's division.

DATE	June 9
BODY	Sun. L.L.
hs	69 - 48.5
IC	- 1.4
D	- 2.9
ha	69 - 44.2
R	+15.6
Ho	69 - 59.8
W	15 - 56-55
corr	0 0
GMT	15 - 56 - 55
gha	45 - 13.2
incr	14 - 13.8
GHA	59 - 27.0
	360
	419 - 27.0
aλ	- 70 - 27.0w
LHA	349
Dec	22 - 56.2N
aL	41 N
Tab Hc	68 - 51
corr	+51
Hc	69 - 42.0
Ho	69 - 59.8
a	17.8T
Zn	151°

Workform for an Observation of the Sun

Before leaving the subject of calculating the altitude and azimuth, I should comment on the use of hand-held calculators, which, while certainly not fundamental to breaking into celestial navigation, do represent a revolutionary potential for performing all kinds of mathematical functions. Although the present rate of development is such that today's miracle becomes tomorrow's routine, there are still underlying considerations which may help you decide what part a calculator should play in your learning and practicing celestial.

Two principal areas in which a hand-held calculator can be useful to the celestial navigator are in deriving the almanac data and in reducing the sight. The latter exercise is quite straightforward in concept, involving the calculator instead of the tables for solving the spherical trig problems. You can use either a budget-priced, scientific type in which you make all the entries by punching the keys and, incidentally, saving very little time and effort as against the tables, or one of the more expensive, programmable types in which the key strokes are pre-programmed and stored on a magnetic tape or chip.

It may soon be practicable to obtain all the almanac data from the more advanced calculators but, at the moment, it is a somewhat cumbersome process and I have not yet found it to save significant time over the almanac itself, especially when I'm in practice.

More important for the small-boat sailor and freshman navigator, are the requirements for energy supply and technical input. Today's generation of small calculators requires that unless you have a supplementary system for recharging the power cells on board every few hours, you are going to have to carry and use up a lot of batteries. Then, if you are going to play any part in developing the formulae or in programming the calculator yourself, this implies a working knowledge of celestial theory, mathematical procedure, and calculator programming and operation. While this is great entertainment for the initiated, the beginner in celestial navigation, unless he acquires "canned" programs and runs them by rote, is likely to be overwhelmed.

Finally, you must recognize that the environment on a small boat is notoriously inhospitable to fine electronic equipment and, for safety's sake, you will always want to have an almanac and sight reduction tables aboard as a back-up. After all, the major advantage of celestial over all other offshore navigation systems is that it is self-contained and dependent upon the navigator alone.

6. Plotting the Line of Position

Your sixth and last step is to plot the line of position resulting from your observation. First, the Assumed Position (AP) is entered on your chart or plotting sheet and so marked. You have it in your workbook: aL 41° N., aλ 70° 27′ W. Next, the intercept (a) is stepped off with your dividers along the azimuth line from the AP. Remembering that a nautical mile is equal to a minute of lati-

Figure 6–1. Plotting the Line of Position from a Sextant Observation. The Intercept is drawn from the Assumed Position toward (or away from, when called for) the direction of the Azimuth. A line is constructed at the intercept point at right angles. This is the Line of Position

POSITION PLOTTING SHEET

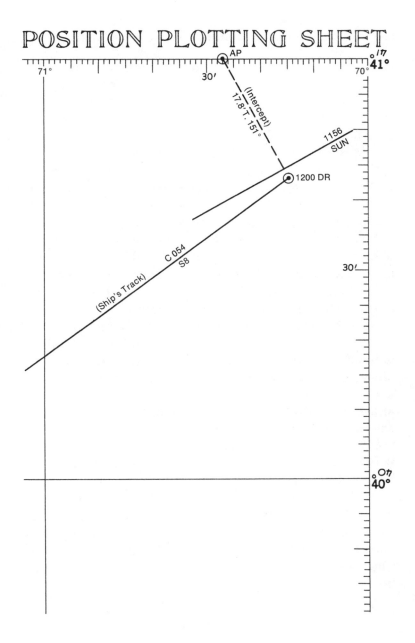

71° 30' 70° 41° °/17

AP

(Intercept)
17.8'T. 151°

1156
SUN

1200 DR

C 054
S8

(Ship's Track)

30'

40° °On

tude, in your example you would step off 17.8 miles along the line *towards* the direction 151° True from your AP. At the end of the intercept construct a line exactly at right angles to the azimuth, and that is your line of position.

The reason the line of position is drawn at right angles to the azimuth is that it is really part of the circumference of a huge circle—a circle so large that a tangent line at the point where its radius intersects the circumference is, for practical purposes, reasonably accurate.

A simple way to construct the plot is to use a small, plastic, draftsman's right triangle, and lay it at the intercept point while holding your protractor or parallel rule along the azimuth line. Figure 6-1 illustrates the plot for the sun line you have worked.

Now that your line of position is established, it can be advanced or "retired" (navigator's lingo for retarded) or crossed with position lines from visual, electronic or other celestial observations to produce your fix.

There are divergent opinions as to whether it is better to use a plotting sheet or a chart for your plot. When I am well offshore, I prefer to use a plotting sheet. I like the 900 series (formerly 3000-Z) Position Plotting Sheets which use a mercator projection like your charts, and have a scale of four inches to 1° of longitude, a convenient dimension for use on a yacht. You could, of course use a chart if you were satisfied that the scale was not too small, and I like to do this when I am nearing land and might have the opportunity to cross my celestial lines with those from terrestrial or electronic observations.

7. The Moon

From time immemorial, the moon has had an undeservedly poor press among practicing navigators. I suspect that, because of its relative nearness and changing size, as well as its irregular motion, sights of the moon and their subsequent reduction complicated the navigator's already burdened existence, and he simply avoided the confrontation. With today's almanac data and your inspection tables for sight reduction, you needn't be timorous.

I have found moon sights to be particularly useful at times, and I think you will too. Quite often, one of the moon's limbs is sufficiently well defined to make a good sextant observation during the daytime, permitting a simultaneous observation with that of the sun, and an immediate fix. A moon shot at twilight, due to the moon's greater luminosity, can often be made while the horizon is still comparatively bright, yielding results superior to those from any of the lesser bodies. It can be added to a round of star or planet sights very beneficially. During World War II, submarine navigators, using sextants to which they had attached eyepieces, like the 7 × 50, with light-gathering ability, took moon sights when they surfaced at night, and even the occasional star sight using the moon-illuminated horizon. If you try this, let me caution you that it is somewhat difficult to separate the true from a false horizon which seems to appear below a low moon, so judge your results accordingly.

You will be pleased to discover that the procedure for working a moon sight is quite similar to that you learned for the sun, with just a few additional steps thrown in. The sextant observation is made in exactly the same way as for the sun, bringing either the lower or upper limb down to tangency with the horizon. The correction for IC and Dip also follows the sun's procedure to obtain ha. But then, to correct ha to Ho, you open to the *back* cover of the almanac where the moon's correction tables (an example of which is shown in Figure 7-1) are located.

Let's walk through a practical example, following the six standard steps from the sextant observation to plotting the line of position, and I'll point out the special peculiarities which apply to the moon.

You're the navigator, approaching Nantucket Shoals on June 9th, 1978, and you are able to make a daytime observation of the moon's upper limb at 15-57-05 GMT.

Figure 7-1. Example of Moon's Altitude Correction Tables from Nautical Almanac

ALTITUDE CORRECTION TABLES 35°–90°—MOON

App. Alt.	35°–39° Corrⁿ	40°–44° Corrⁿ	45°–49° Corrⁿ	50°–54° Corrⁿ	55°–59° Corrⁿ	60°–64° Corrⁿ	65°–69° Corrⁿ	70°–74° Corrⁿ	75°–79° Corrⁿ	80°–84° Corrⁿ	85°–89° Corrⁿ	App. Alt.
00	35 56·5	40 53·7	45 50·5	50 46·9	55 43·1	60 38·9	65 34·6	70 30·1	75 25·3	80 20·5	85 15·6	00
10	56·4	53·6	50·4	46·8	42·9	38·8	34·4	29·9	25·2	20·4	15·5	10
20	56·3	53·5	50·2	46·7	42·8	38·7	34·3	29·7	25·0	20·2	15·3	20
30	56·2	53·4	50·1	46·5	42·7	38·5	34·1	29·6	24·9	20·0	15·1	30
40	56·2	53·3	50·0	46·4	42·5	38·4	34·0	29·4	24·7	19·9	15·0	40
50	56·1	53·2	49·9	46·3	42·4	38·2	33·8	29·3	24·5	19·7	14·8	50
00	36 56·0	41 53·1	46 49·8	51 46·2	56 42·3	61 38·1	66 33·7	71 29·1	76 24·4	81 19·6	86 14·6	00
10	55·9	53·0	49·7	46·0	42·1	37·9	33·5	29·0	24·2	19·4	14·5	10
20	55·8	52·8	49·5	45·9	42·0	37·8	33·4	28·8	24·1	19·2	14·3	20
30	55·7	52·7	49·4	45·8	41·8	37·7	33·2	28·7	23·9	19·1	14·1	30
40	55·6	52·6	49·3	45·7	41·7	37·5	33·1	28·5	23·8	18·9	14·0	40
50	55·5	52·5	49·2	45·5	41·6	37·4	32·9	28·3	23·6	18·7	13·8	50
00	37 55·4	42 52·4	47 49·1	52 45·4	57 41·4	62 37·2	67 32·8	72 28·2	77 23·4	82 18·6	87 13·7	00
10	55·3	52·3	49·0	45·3	41·3	37·1	32·6	28·0	23·3	18·4	13·5	10
20	55·2	52·2	48·8	45·2	41·2	36·9	32·5	27·9	23·1	18·2	13·3	20
30	55·1	52·1	48·7	45·0	41·0	36·8	32·3	27·7	22·9	18·1	13·2	30
40	55·0	52·0	48·6	44·9	40·9	36·6	32·2	27·6	22·8	17·9	13·0	40
50	55·0	51·9	48·5	44·8	40·8	36·5	32·0	27·4	22·6	17·8	12·8	50
00	38 54·9	43 51·8	48 48·4	53 44·6	58 40·6	63 36·4	68 31·9	73 27·2	78 22·5	83 17·6	88 12·7	00
10	54·8	51·7	48·2	44·5	40·5	36·2	31·7	27·1	22·3	17·4	12·5	10
20	54·7	51·6	48·1	44·4	40·3	36·1	31·6	26·9	22·1	17·3	12·3	20
30	54·6	51·5	48·0	44·2	40·2	35·9	31·4	26·8	22·0	17·1	12·2	30
40	54·5	51·4	47·9	44·1	40·1	35·8	31·3	26·6	21·8	16·9	12·0	40
50	54·4	51·2	47·8	44·0	39·9	35·6	31·1	26·5	21·7	16·8	11·8	50
00	39 54·3	44 51·1	49 47·6	54 43·9	59 39·8	64 35·5	69 31·0	74 26·3	79 21·5	84 16·6	89 11·7	00
10	54·2	51·0	47·5	43·7	39·6	35·3	30·8	26·1	21·3	16·5	11·5	10
20	54·1	50·9	47·4	43·6	39·5	35·2	30·7	26·0	21·2	16·3	11·4	20
30	54·0	50·8	47·3	43·5	39·4	35·0	30·5	25·8	21·0	16·1	11·2	30
40	53·9	50·7	47·2	43·3	39·2	34·9	30·4	25·7	20·9	16·0	11·0	40
50	53·8	50·6	47·0	43·2	39·1	34·7	30·2	25·5	20·7	15·8	10·9	50

H.P.	L U	L U	L U	L U	L U	L U	L U	L U	L U	L U	L U	H.P.
54·0	1·1 1·7	1·3 1·9	1·5 2·1	1·7 2·4	2·0 2·6	2·3 2·9	2·6 3·2	2·9 3·5	3·2 3·8	3·5 4·1	3·8 4·5	54·0
54·3	1·4 1·8	1·6 2·0	1·8 2·2	2·0 2·5	2·3 2·7	2·5 3·0	2·8 3·2	3·0 3·5	3·3 3·8	3·6 4·1	3·9 4·4	54·3
54·6	1·7 2·0	1·9 2·2	2·1 2·4	2·3 2·6	2·5 2·8	2·7 3·0	3·0 3·3	3·2 3·5	3·5 3·8	3·7 4·1	4·0 4·3	54·6
54·9	2·0 2·2	2·2 2·3	2·3 2·5	2·5 2·7	2·7 2·9	2·9 3·1	3·2 3·3	3·4 3·5	3·6 3·8	3·9 4·0	4·1 4·3	54·9
55·2	2·3 2·3	2·5 2·4	2·6 2·6	2·8 2·8	3·0 2·9	3·2 3·1	3·4 3·3	3·6 3·5	3·8 3·7	4·0 4·0	4·2 4·2	55·2
55·5	2·7 2·5	2·8 2·6	2·9 2·7	3·1 2·9	3·2 3·0	3·4 3·2	3·6 3·4	3·7 3·5	3·9 3·7	4·1 3·9	4·3 4·1	55·5
55·8	3·0 2·6	3·1 2·7	3·2 2·8	3·3 3·0	3·5 3·1	3·6 3·3	3·8 3·4	3·9 3·6	4·1 3·7	4·2 3·9	4·4 4·0	55·8
56·1	3·3 2·8	3·4 2·9	3·5 3·0	3·6 3·1	3·7 3·2	3·8 3·3	4·0 3·4	4·1 3·6	4·2 3·7	4·4 3·8	4·5 4·0	56·1
56·4	3·6 2·9	3·7 3·0	3·8 3·1	3·9 3·2	3·9 3·3	4·0 3·4	4·1 3·5	4·3 3·6	4·4 3·7	4·5 3·8	4·6 3·9	56·4
56·7	3·9 3·1	4·0 3·1	4·1 3·2	4·1 3·3	4·2 3·3	4·3 3·4	4·3 3·4	4·4 3·5	4·5 3·7	4·6 3·7	4·7 3·8	56·7
57·0	4·3 3·2	4·3 3·3	4·3 3·3	4·4 3·4	4·4 3·4	4·5 3·5	4·5 3·5	4·6 3·6	4·7 3·6	4·7 3·7	4·8 3·8	57·0
57·3	4·6 3·4	4·6 3·4	4·6 3·4	4·6 3·5	4·7 3·5	4·7 3·5	4·7 3·6	4·8 3·6	4·8 3·6	4·8 3·7	4·9 3·7	57·3
57·6	4·9 3·6	4·9 3·6	4·9 3·6	4·9 3·6	4·9 3·6	4·9 3·6	4·9 3·6	5·0 3·6	5·0 3·6	5·0 3·6	5·0 3·6	57·6
57·9	5·2 3·7	5·2 3·7	5·2 3·7	5·2 3·7	5·2 3·7	5·1 3·6	5·1 3·6	5·1 3·6	5·1 3·6	5·1 3·6	5·1 3·6	57·9
58·2	5·5 3·9	5·5 3·8	5·5 3·8	5·4 3·8	5·4 3·7	5·4 3·7	5·3 3·7	5·3 3·6	5·2 3·6	5·2 3·5	5·2 3·5	58·2
58·5	5·9 4·0	5·8 4·0	5·8 3·9	5·7 3·9	5·6 3·8	5·6 3·8	5·5 3·7	5·5 3·6	5·4 3·5	5·3 3·5	5·3 3·4	58·5
58·8	6·2 4·2	6·1 4·1	6·0 4·1	6·0 4·0	5·9 3·9	5·8 3·8	5·7 3·7	5·6 3·5	5·5 3·5	5·4 3·5	5·3 3·4	58·8
59·1	6·5 4·3	6·4 4·3	6·3 4·2	6·2 4·1	6·1 4·0	6·0 3·9	5·9 3·8	5·8 3·6	5·7 3·5	5·6 3·4	5·4 3·3	59·1
59·4	6·8 4·5	6·7 4·4	6·6 4·3	6·5 4·2	6·4 4·1	6·2 3·9	6·1 3·8	6·0 3·7	5·8 3·5	5·7 3·4	5·5 3·2	59·4
59·7	7·1 4·6	7·0 4·5	6·9 4·4	6·8 4·3	6·6 4·1	6·5 4·0	6·3 3·8	6·2 3·7	6·0 3·5	5·8 3·3	5·6 3·2	59·7
60·0	7·5 4·8	7·3 4·7	7·2 4·5	7·0 4·4	6·9 4·2	6·7 4·0	6·5 3·9	6·3 3·7	6·1 3·5	5·9 3·3	5·7 3·1	60·0
60·3	7·8 5·0	7·6 4·8	7·5 4·7	7·3 4·5	7·1 4·3	6·9 4·1	6·7 3·9	6·5 3·7	6·3 3·5	6·0 3·2	5·8 3·0	60·3
60·6	8·1 5·1	7·9 5·0	7·7 4·8	7·6 4·6	7·3 4·4	7·1 4·2	6·9 3·9	6·7 3·7	6·4 3·4	6·2 3·2	5·9 2·9	60·6
60·9	8·4 5·3	8·2 5·1	8·0 4·9	7·8 4·7	7·6 4·5	7·3 4·2	7·1 4·0	6·8 3·7	6·6 3·4	6·3 3·2	6·0 2·9	60·9
61·2	8·7 5·4	8·5 5·2	8·3 5·0	8·1 4·8	7·8 4·5	7·6 4·3	7·3 4·0	7·0 3·7	6·7 3·4	6·4 3·1	6·1 2·8	61·2
61·5	9·1 5·6	8·8 5·4	8·6 5·1	8·3 4·9	8·1 4·6	7·8 4·3	7·5 4·0	7·2 3·7	6·9 3·4	6·5 3·1	6·2 2·7	61·5

HP

The sextant altitude reads 36-05.3 and your instrument has an Index Correction of −1.4. Your height of eye is nine feet. Your dead reckoning position (DR) at the approximate time of the sight is Latitude 40°51′ North, Longitude 69° 47′ West. You chose the upper limb in this instance because it was clearly defined and you could get a better sight than with the lower limb.

Being a firm believer in the navigator's workbook, I have shown a suggested workform for moon sights, filled in with the figures for this sight. The steps in completing it, which you can follow with the text and tabular excerpts in this chapter, start with the altitude corrections from the back of the almanac. Figure 7-2 shows the portion applicable to our sight.

The IC and Dip are found, and entered onto the workform just as you did with the sun, resulting, in this example in an ha of 36-01.0. Then, to correct ha to Ho, you enter the main body of the moon table which, you will notice, presents the correction in two parts. The first, or R correction is taken from the upper box in which you will find the *degrees* of your apparent altitude (ha) in one of the vertical columns. Descending vertically to a point opposite the *minutes* of your ha, which are shown in ten minute increments in the column at the side of the table, you can extract the first correction, interpolating, if necessary, by eye. In our example, you descend the 35°-39° column to a point opposite 36-00, and find the first correction of 56.0′. A quick inspection tells you that the correction is changing slowly at that altitude and the 56.0 reading applies equally to the 36-01 ha.

The second correction, for horizontal parallax (H.P.), which applies to the moon alone, is found by descending the same column as for the first correction into the

ALTITUDE CORRECTION TABLES
35°–90°—MOON

App. Alt.	35°–39° Corrⁿ	40°–44° Corrⁿ	45°–49° Corrⁿ	50°–54° Corrⁿ	DIP Ht. of Eye	Corrⁿ	Ht. of Eye	Ht. of Eye	Corrⁿ	Ht. of Eye
00	35 56·5	40 53·7	45 50·5	50 46·9						
10	56·4	53·6	50·4	46·8	m		ft.	m		ft.
20	56·3	53·5	50·2	46·7	2·4	−2·8	8·0	9·5	−5·5	31·5
30	56·2	53·4	50·1	46·5	2·6		8·6	9·9		32·7
40	56·2	53·3	50·0	46·4	2·8	−2·9	9·2	10·3	−5·6	33·9
50	56·1	53·2	49·9	46·3		−3·0			−5·7	
00	36 56·0	41 53·1	46 49·8	51 46·2	3·0	−3·1	9·8	10·6	−5·8	35·1
10	55·9	53·0	49·7	46·0	3·2	−3·2	10·5	11·0	−5·9	36·3
20	55·8	52·8	49·5	45·9	3·4	−3·3	11·2	11·4	−6·0	37·6
30	54·4	51·2	49·4	44·0	3·6	−3·4	11·9	11·8	−6·1	38·9
00	39 54·3	44 51·1	49 47·6	54 43·9	3·8	−3·5	12·6	12·2	−6·2	40·1
10	54·2	51·0	47·5	43·7	4·0	−3·6	13·3	12·6	−6·3	41·5
20	54·1	50·9	47·4	43·6	4·3	−3·7	14·1	13·0	−6·4	42·8
30	54·0	50·8	47·3	43·5	4·5	−3·8	14·9	13·4	−6·5	44·2
40	53·9	50·7	47·2	43·3	4·7	−3·9	15·7	13·8	−6·6	45·5
50	53·8	50·6	47·0	43·2	5·0	−4·0	16·5	14·2	−6·7	46·9
					5·2	−4·1	17·4	14·7	−6·8	48·4
H.P.	L U	L U	L U	L U	5·5	−4·2	18·3	15·1	−6·9	49·8
					5·8	−4·3	19·1	15·5	−7·0	51·3
54·0	1·1 1·7	1·3 1·9	1·5 2·1	1·7 2·4	6·1	−4·4	20·1	16·0	−7·1	52·8
54·3	1·4 1·8	1·6 2·0	1·8 2·2	2·0 2·5	6·3	−4·5	21·0	16·5	−7·2	54·3
54·6	1·7 2·0	1·9 2·2	2·1 2·4	2·3 2·6	6·6	−4·6	22·0	16·9	−7·3	55·8
54·9	2·0 2·2	2·2 2·3	2·3 2·5	2·5 2·7	6·9	−4·7	22·9	17·4	−7·4	57·4
55·2	2·3 2·3	2·5 2·4	2·6 2·6	2·8 2·8	7·2		22·0	17·9	−7·5	58·9
55·5	2·7									

...... part for lower (L) and upper (U) limbs. All corrections are to be **added** to apparent altitude, *but 30′ is to be subtracted from the altitude of the upper limb*,

...titude.

App. Alt. = Apparent altitude = Sextant altitude corrected for index error and dip.

Figure 7–2. Excerpt from Moon's Altitude Correction Tables from Nautical Almanac showing Dip correction (−2.9′ for Ht. of Eye of 9 ft.), Refraction correction (56.0′ for App. Alt. of 36° 01.0′) and Horizontal Parallax correction (1.7′ for H.P. of 54.0′, Upper Limb)

lower box, and, under the heading "U" for upper limb (obviously you would use heading "L" had you shot the lower limb), extract the value of the second correction opposite the value for H.P. you will have found in the daily pages of the almanac as will be described momentarily. Remember that both the first (R) and second (H.P.) corrections are always *added* to the moon's apparent altitude and, in the case of an observation of the *upper* limb, a third value, invariably exactly 30 minutes, is *subtracted* from the total. In our example, given an H.P. of 54.0, the second correction is 1.7', which, when added to ha and the first correction, with 30' being subtracted for the upper limb, results in an Ho of 36-28.7

Now you're wondering, quite reasonably, what horizontal parallax is and where you obtain the value of it to use in the lower box of the correction table, so I'll keep the secret no longer. Horizontal parallax (H.P.) is the difference in altitude between that measured from the observer's position on the surface of the earth and that measured from the center of the earth. It is most important in moon observations because of the relatively close distance of the moon from the earth. As to the value of H.P. for any given time, when you open the almanac to the daily pages to find the GHA and Dec data (Figure 7-3), you will discover in the moon's fifth column, opposite the nearest hour of GMT, the H.P. you will use for the second altitude correction (54.0').

Having corrected the sight to obtain your Ho, the next move is to determine the moon's GHA for the time of the sight. The almanac (Figure 7-3) is entered in the same way as for the sun, except, of course, you use the MOON columns. In our example, the uncorrected Greenwich

1978 JUNE 9, 10, 11 (FRI., SAT., SUN.)

G.M.T.	SUN		MOON				
	G.H.A.	Dec.	G.H.A.	v	Dec.	d	H.P.
d h	° ′	° ′	° ′	′	° ′	′	′
➤9 00	180 15.0	N22 52.8	142 58.0	12.9	N16 50.0	3.9	54.0
01	195 14.9	53.1	157 29.9	12.9	16 46.1	4.0	54.0
02	210 14.8	53.3	172 01.8	12.9	16 42.1	4.1	54.0
03	225 14.7 ··	53.5	186 33.7	13.0	16 38.0	4.2	54.0
04	240 14.6	53.7	201 05.7	13.0	16 33.8	4.2	54.0
05	255 14.4	53.9	215 37.7	13.0	16 29.6	4.3	54.0
06	270 14.3	N22 54.1	230 09.7	13.0	N16 25.3	4.4	54.0
07	285 14.2	54.4	244 41.7	13.1	16 20.9	4.5	54.0
08	300 14.1	54.6	259 13.8	13.1	16 16.4	4.5	54.0
F 09	315 13.9 ··	54.8	273 45.9	13.1	16 11.9	4.6	54.0
R 10	330 13.8	55.0	288 18.0	13.1	16 07.3	4.7	54.0
I 11	345 13.7	55.2	302 50.1	13.1	16 02.6	4.7	54.0
D 12	0 13.6	N22 55.4	317 22.2	13.2	N15 57.9	4.8	54.0
A 13	15 13.5	55.6	331 54.4	13.2	15 53.1	4.9	54.0
Y 14	30 13.3	55.8	346 26.6	13.2	15 48.2	5.0	54.0
15	45 13.2 ··	56.0	0 58.8	13.2	15 43.2	5.0	54.0
16	60 13.1	56.2	15 31.0	13.3	15 38.2	5.1	54.0
17	75 13.0	56.5	30 03.3	13.3	15 33.1	5.2	54.0
18	90 12.8	N22 56.7	44 35.6	13.3	N15 27.9	5.2	54.1
19	105 12.7	56.9	59 07.9	13.3	15 22.7	5.3	54.1
20	120 12.6	57.1	73 40.2	13.3	15 17.4	5.3	54.1
21	135 12.5 ··	57.3	88 12.5	13.4	15 12.1	5.5	54.1
22	150 12.4	57.5	102 44.9	13.4	15 06.6	5.5	54.1
23		57.7	117 17.3	13.4	15 01.1	5.5	54.1

Figure 7–3. Excerpt from Nautical Almanac showing Astronomical Data for the Moon on June 9 at 15ʰ G.M.T.

Hour Angle, 0-58.8, is found opposite the appropriate hour of GMT (15ʰ in this case), while the increment for 57 minutes and 5 seconds, 13-37.2, is found in the yellow pages in the back of the almanac (Figure 7-4). Be careful to use the MOON column.

NAVIGATOR'S WORKBOOK

DATE	June 9
BODY	Moon, U.L.
hs	36 - 05.3
IC	- 1.4
D	- 2.9
ha	36 - 01.0
R	+ 56.0
H.P. [H.P.]	+ 1.7 [54.0]
(−30′)	-30
Ho	36 - 28.7
W	15 -57-05
corr	00
GMT	15 - 57 - 05
gha [v]	0 - 58.8 [12.2]
incr	13 - 37.2
v corr	12.7
GHA	14 - 48.7
	+360
	374 - 48.7
aλ	69 - 48.7 W
LHA	305
dec [d]	15 -43.2 N [5.0]
d corr	- 4.8
Dec	15 - 38.4 N
aL	41 N
Tab Hc	36 - 01
corr	+ 24
Hc	36 - 25.0
Ho	36 - 28.7
a	3.7 T
Zn	102

Workform for an Observation of the Moon

INCREMENTS AND CORRECTIONS 57ᵐ

57ᵐ	SUN PLANETS	ARIES	MOON	v or Corrⁿ d		v or Corrⁿ d		v or Corrⁿ d	
s	° ′	° ′	° ′	′	′	′	′	′	′
00	14 15·0	14 17·3	13 36·1	0·0	0·0	6·0	5·8	12·0	11·5
01	14 15·3	14 17·6	13 36·3	0·1	0·1	6·1	5·8	12·1	11·6
02	14 15·5	14 17·8	13 36·5	0·2	0·2	6·2	5·9	12·2	11·7
03	14 15·8	14 18·1	13 36·8	0·3	0·3	6·3	6·0	12·3	11·8
04	14 16·0	14 18·3	13 37·0	0·4	0·4	6·4	6·1	12·4	11·9
05	14 16·3	14 18·6	13 37·2	0·5	0·5	6·5	6·2	12·5	12·0
06	14 16·5	14 18·8	13 37·5	0·6	0·6	6·6	6·3	12·6	12·1
07	14 16·8	14 19·1	13 37·7	0·7	0·7	6·7	6·4	12·7	12·2
08	14 17·0	14 19·3	13 38·0	0·8	0·8	6·8	6·5	12·8	12·3
09	14 17·3	14 19·6	13 38·2	0·9	0·9	6·9	6·6	12·9	12·4
10	14 17·5	14 19·8	13 38·4	1·0	1·0	7·0	6·7	13·0	12·5
11	14 17·8	14 20·1	13 38·7	1·1	1·1	7·1	6·8	13·1	12·6
12	14 18·0	14 20·3	13 38·9	1·2	1·2	7·2	6·9	13·2	12·7
13	14 18·3	14 20·6	13 39·2	1·3	1·2	7·3	7·0	13·3	12·7
14	14 ·	· · 39·4		1·4	1·3	7· ·		· · ·
49	14 27·3	14 29·6	13 4 /·/	4·9	4·7	10·9	10·4	16·9	16·2
50	14 27·5	14 29·9	13 48·0	5·0	4·8	11·0	10·5	17·0	16·3
51	14 27·8	14 30·1	13 48·2	5·1	4·9	11·1	10·6	17·1	16·4
52	14 28·0	14 30·4	13 48·5	5·2	5·0	11·2	10·7	17·2	16·5
53	14 28·3	14 30·6	13 48·7	5·3	5·1	11·3	10·8	17·3	16·6
54	14 28·5	14 30·9	13 48·9	5·4	5·2	11·4	10·9	17·4	16·7
55	14 28·8	14 31·1	13 49·2	5·5	5·3	11·5	11·0	17·5	16·8
56	14 29·0	14 31·4	13 49·4	5·6	5·4	11·6	11·1	17·6	16·9
57	14 29·3	14 31·6	13 49·7	5·7	5·5	11·7	11·2	17·7	17·0
58	14 29·5	14 31·9	13 49·9	5·8	5·6	11·8	11·3	17·8	17·1
59	14 29·8	14 32·1	13 50·1	5·9	5·7	11·9	11·4	17·9	17·2
60	14 30·0	14 32·4	13 50·4	6·0	5·8	12·0	11·5	18·0	17·3

Figure 7–4. Excerpt from Nautical Almanac Increments and Corrections Tables showing Moon's increment for 57ᵐ 05ˢ and v and d corrections

In the case of the moon, there is a small, additional increment called the "v" correction which recognizes the excesses in the moon's actual movement over the constant rate used in the increment table. The value of v is found in the second column of moon data (Figure 7-3) next to the GHA you extracted. Going to that v value in the right hand columns of the Increments and Corrections table (Figure 7-4), take out the v correction (in our example, a correction of 12.7' for a v of 13.2), and add it to your increment for minutes and seconds of time to arrive at the correct total GHA.

Declination is found in the third of the moon's columns on the daily page (Figure 7-3) opposite the GMT hour of the observation. While this Dec figure can be adjusted for the minutes and seconds of time by inspection, as you did with the sun, it is probably quicker and more accurate to note the value for d (which represents the hourly change in declination) in the fourth column of the moon table (Figure 7-3), and take it to the Increments table (Figure 7-4) where the d correction is found in the same manner as you found the v increment. In our example, the d correction for a d of 5.0 is 4.8', which, when applied to the tabular declination, produces a total Dec of 15-38.4 N. Note that while the v increment is always added, it is necessary to check whether the declination is increasing or decreasing (in our case it was the latter), so that you apply the d correction with the proper sign.

The "special" treatment required of a moon sight is now completed. For clarity, I have described each step in some detail, but you will find that with your workform and a little practice, the few extra steps involved with moon shots are really very routine.

Having obtained GHA, you apply your Assumed Longitude (selecting it to make LHA come out to a whole degree) and arrive at LHA, in this example, 305, which, together with your Dec and an Assumed Latitude near

Figure 7-5. Excerpt from Table of Computed Altitudes and Azimuths, Pub. No. 249, Vol III, for Latitude 41°, Declination 15°—Same Name as Latitude—and Local Hour Angle, 305

N. Lat. {LHA greater than 180°......... Zn=Z
{LHA less than 180°...........Zn=360−Z

LAT 41°
DECLINATION (15°-29°) SAME NAME AS LATITUDE

	15°			16°			17°			18°			19°			
LHA	Hc	d	Z	Hc	d	Z	Hc	d	Z	Hc	d	Z	Hc	d	Z	LHA
	o ′	′	o	o ′	′	o	o ′	′	o	o ′	′	o	o ′	′	o	
0	64 00 +60		180	65 00 +60		180	66 00 +60		180	67 00 +60		180	68 00 +60		180	360
1	63 59	60	178	64 59	60	178	65 59	60	178	66 59	60	178	67 59	60	178	359
2	63 57	59	176	64 56	60	176	65 56	60	175	66 56	60	175	67 56	60	175	358
3	63 52	60	173	64 52	60	173	65 52	59	173	66 51	60	173	67 51	60	173	357
4	63 ᵈᶜ		171			171	65 45	60	171					60	170	35ᶜ
52	38 13	39	104	38 52	38	102			102	40 08	38	101	40 46	37	100	308
53	37 29	39	104	38 08	38	103	38 46	38	102	39 24	37	101	40 01	37	100	307
54	36 45	39	103	37 24	38	102	38 02	37	101	38 39	37	100	39 16	37	99	306
55	36 01 +38		102	36 39 +38		101	37 17 +38		100	37 55 +37		99	38 32 +36		98	305
56	35 16	39	101	35 55	38	100	36 33	37	99	37 10	37	98	37 47	36	97	304
57	34 32	38	101	35 10	38	100	35 48	37	99	36 25	37	98	37 02	36	97	303
58	33 47	38	100	34 25	38	99	35 03	37	98	35 40	37	97	36 17	36	96	302
59	33 03	38	99	33 41	37	98	34 18	37	97	34 55	37	96	35 32	36	95	301
60	32 18 +38		98	32 56 +37		97	33 33 +37		96	34 10 +37		96	34 47 +36		95	300
61	31 33	38	98	32 11	37	97	32 48	37	96	33 25	37	95	34 02	36	94	299
62	30 48	38	97	31 26	37	96	32 03	37	95	32 40	36	94	33 16	36	93	298
63	30 03	38	96	30 41	37	95	31 18	37	94	31 55	36	93	32 31	36	92	297
64	29 18	38	95	29 56	37	95	30 33	36	94	31 09	37	93	31 46	36	92	296
65	28 33 +37		95	29 10 +37		94	29 47 +37		93	30 24 +37		92	31 01 +36		91	295
66	27 48	37	94	28 25	37	93	29 02	37	92	29 39	36	91	30 15	36	90	294
67	27 03	37	93	27 40	37	93	28 17	37	92	28 54	36	91	29 30	36	90	293
68	26 17	38	93	26 55	37	92	27 32	36	91	28 08	37	90	28 45	36	89	292
69	25 32	37	92	26 09	37	91	26 46	37	90	27 23	36	89	27 59	36	89	291
	15°			16°			17°			18°			19°			

your DR (41°N), provide all the information necessary to proceed to the sight reduction table. Figure 7-5 is an excerpt from the appropriate table in Pub. No. 249, Vol. III.

Following the procedure you learned for the sun, you enter the column for 15° declination, *same* name as your Assumed Latitude of 41° N, and take out, opposite your LHA of 305, the tabular Hc of 36-01, the value of *d* (+38), and the uncorrected azimuth (Z) of 102°.

The tabular value for Hc needs to be corrected for the incremental minutes of declination (38.4) which is accomplished with Table 5, in the back of Pub. No. 249, an excerpt of which is shown in Figure 7-6.

In our example, the altitude differential, *d*, of +38', applied to the remaining, incremental minutes of declination, 38.4, yields a correction of 24 minutes which is *added* to the tabular Hc to obtain the corrected Hc because the differential, *d*, carries a positive sign.

The corrected Hc is next compared with the Observed Altitude (Ho) obtained earlier, to arrive at the altitude difference or intercept. In our example, Ho is greater, so the intercept of 3.7' is "Towards". Applying the rules in the upper corner of the sight reduction table (Figure 7-5), we see that Zn is equal, in our case, to Z, so our True Azimuth is 102°.

Your final step is to plot the intercept from your Assumed Position in the direction of the azimuth and, at that point, construct your line of position at right angles to the azimuth exactly as you have done with the sun line.

Try the moon when you get the opportunity. I'm sure you will find the satisfaction well worth the few extra steps this valuable body requires.

TABLE 5.—Correction to Tabulated Altitude

34 35 36	37 38 39	40 41 42	4	55 56 57	58 59 60	d/
0 0 0	0 0 0	0 0 0		0 0 0	0 0 0	0
1 1 1	1 1 1	1 1 1		1 1 1	1 1 1	1
1 1 1	1 1 1	1 1 1	1	2 2 2	2 2 2	2
2 2 2	2 2 2	2 2 2	2	3 3 3	3 3 3	3
2 2 2	2 3 3	3 3 3	3	4 4 4	4 4 4	4
3 3 3	3 3 3	3 3 4	4	5 5 5	5 5 5	5
3 4 4	4 4 4	4 4 4	4	6 6 6	6 6 6	6
4 4 4	4 4 5	5 5 5	5	6 7 7	7 7 7	7
5 5 5	5 5 5	5 5 6	6	7 7 8	8 8 8	8
5 5				8 8 9		
16 16 17	17 18 18	19 19 20	2	26 26 27	27 28 28	28
16 17 17	18 18 19	19 20 20	2	27 27 28	28 29 29	29
17 18 18	18 19 20	20 20 21	2	28 28 28	29 30 30	30
8 18 19	19 20 20	21 21 22	2	28 29 29	30 30 31	31
8 19 19	20 20 21	21 22 22	2	29 30 30	31 31 32	32
9 19 20	20 21 21	22 23 23	2	30 31 31	32 32 33	33
9 20 20	21 22 22	23 23 24	24	31 32 32	33 33 34	34
0 20 21	22 22 23	23 24 24	25	32 33 33	34 34 35	35
0 21 22	22 23 23	24 25 25	26	33 34 34	35 35 36	36
1 22 22	23 23 24	25 25 26	27	34 35 35	36 36 37	37
22 22 23	23 24 25	25 26 27	27	35 35 36	37 37 38	38
22 23 23	24 25 25	26 27 27	2	36 36 37	38 38 39	39
23 23 24	25 25 26	27 27 28	2	37 37 38	39 39 40	40
23 24				38 38		

Figure 7–6. Excerpt from Table 5, Pub. No. 249, showing Correction to Tabulated Altitude for d of 38, and Incremental Minutes, 38

8. The Planets

Four planets are of particular interest to celestial navigators: Venus, Mars, Jupiter and Saturn. Of these, Venus is the most useful. Not only is it the brightest body in the heavens besides the sun and moon, but it is also easy to identify as a morning or evening "star," not far from the sun, and, under ideal conditions, can be seen sufficiently well in the daytime to make an observation.

Jupiter is also quite brilliant, often being brighter than Sirius, the brightest star in the skies. Jupiter can frequently provide good observations at morning or evening twilight. I have found Mars and Saturn much more difficult to distinguish from the stars while there is still enough light to see the horizon clearly. When planning to observe either of the last two, I usually work through a quick, trial reduction, using roughly approximated inputs, to estimate an altitude and azimuth so I'll know where to look for them.

In observing the planets with a sextant, you must be sure to bring the *center* of the body to the horizon. Unlike the stars which produce only a pinpoint of light, the planets will exhibit a visible disc when viewed through a magnifying eyepiece, and it is important that the center of the disc, and not one of the limbs, be bisected by the horizon.

Let's walk quickly through a practical example, as we have done with the sun and the moon, to review the steps, and to familiarize you with the minor differences which apply to planet sights. The workform shown is suitable for all planet observations and has been completed for this practical example so that you can follow the procedure step by step.

At sunset on June 11th, 1978, in DR position Latitude 40° 41' North, Longitude 72° 21' West, you make an observation of Venus. The GMT is 23-56-14, the sextant altitude (hs) is 30-22.0, the Index Correction is −1.4 and your height of eye is nine feet.

First, as usual, you will apply your IC and Dip corrections to hs to obtain ha in the same way as for other sights. The R correction to ha, to obtain Ho is, however, slightly different. It will be found in the STARS AND PLANETS table inside the front cover of the almanac (Figure 8-1). Note that it is always negative.

Certain named planets may require small, additional corrections during times of the year, and these are listed in the right hand column of the planet table, naturally being applied according to their sign. In the case of your June 11th observation of Venus, you will see that a *plus*

ALTITUDE CORRECTION TABLES 10°-90°

STARS AND PLANETS		DIP		
App. Alt. Corrⁿ	App. Alt. Additional Corrⁿ	Ht. of Eye Corrⁿ	Ht. of Eye	Ht. of Eye Corrⁿ

App. Alt.	Corrⁿ	Additional Corrⁿ	Ht. of Eye (m)	Corrⁿ	Ht. of Eye (ft)	Ht. of Eye (m)	Corrⁿ
° ′		1978	m		ft.	m	′
9 56	-5.3	**VENUS**	2.4	-2.8	8.0	1.0 — 1.8	
10 08	-5.2	Jan. 1-July 20	2.6	-2.9	8.6	1.5 — 2.2	
10 20	-5.1	°	2.8	-3.0	9.2	2.0 — 2.5	
10 33	-5.0	42 + 0′.1	3.0	-3.1	9.8	2.5 — 2.8	
10 46	-4.9		3.2	-3.2	10.5	3.0 — 3.0	
11 00	-4.8	July 21-Sept. 2	3.4	-3.3	11.2	See table	
11 14		° ± 0′.2	3.6	-3.4	11.9	←	
			3.8				

App. Alt.	Corrⁿ	Additional Corrⁿ	Ht. of Eye (m)	Corrⁿ	Ht. of Eye (ft)	Ht. of Eye	Corrⁿ
27 36	-1.9	41	12.6	-6.3	41.5	70	8.1
28 56	-1.8	Dec. 20-Dec 31	13.0	-6.4	42.8	75	8.4
30 24	-1.7	°	13.4	-6.5	44.2	80 — 8.7	
32 00	-1.6	46 + 0′.3	13.8	-6.6	45.5	85	8.9
33 45	-1.5		14.2	-6.7	46.9	90	9.2
35 40	-1.4	**MARS**	14.7	-6.8	48.4	95 — 9.5	
37 48	-1.3	Jan. 1-Mar. 22	15.1	-6.9	49.8		
40 08	-1.2	° + 0′.2	15.5	-7.0	51.3	100 — 9.7	
			16.0				

App. Alt.   Apparent altitude

Sextant altitude corrected for index error and dip.

Figure 8–1. Portion of Altitude Correction Tables from Nautical Almanac showing Dip correction (−2.9' for Ht. of Eye of 9 ft.), Refraction correction (−1.7' for App. Alt. of 30° 17.7') and Special correction for Venus (+0.1' for Alt. 0°–42°)

Figure 8–2. Example of Daily Page for Planets and Stars from Nautical Almanac

G.M.T.	ARIES G.H.A.	VENUS −3.4 G.H.A.	Dec.	MARS +1.5 G.H.A.	Dec.	JUPITER −1.4 G.H.A.	Dec.	SATURN +0.8 G.H.A.	Dec.	STARS Name	S.H.A.	Dec.
d h	° '	° '	° '	° '	° '	° '	° '	° '	° '		° '	° '
9 00	257 00.5	143 33.0 N23 35.7		107 07.1 N13 39.8		155 13.4 N23 05.6		108 53.7 N14 27.4		Acamar	315 38.8 S40 23.4
01	272 03.0	158 32.3	35.3	122 08.3	39.3	170 15.3	05.6	123 56.0	27.4	Achernar	335 46.9 S57 20.6
02	287 05.5	173 31.6	34.9	137 09.4	38.8	185 17.1	05.6	138 58.3	27.3	Acrux	173 38.6 S62 59.1
03	302 07.9	188 30.8 ··	34.4	152 10.6 ··	38.3	200 19.0 ··	05.5	154 00.6 ··	27.2	Adhara	255 33.7 S28 56.8
04	317 10.4	203 30.1	34.0	167 11.8	37.8	215 20.9	05.5	169 02.9	27.2	Aldebaran	291 20.1 N16 27.8
05	332 12.9	218 29.4	33.5	182 13.0	37.3	230 22.8	05.4	184 05.2	27.1			
06	347 15.3	233 28.6 N23 33.1		197 14.2 N13 36.7		245 24.7 N23 05.4		199 07.5 N14 27.0		Alioth	166 43.7 N56 04.9
07	2 17.8	248 27.9	32.7	212 15.3	36.2	260 26.5	05.4	214 09.8	27.0	Alkaid	153 19.5 N49 25.5
08	17 20.3	263 27.2	32.2	227 16.5	35.7	275 28.4	05.3	229 12.0	26.9	Al Na'ir	28 16.8 S47 03.7
F 09	32 22.7	278 26.4 ··	31.8	242 17.7 ··	35.2	290 30.3 ··	05.3	244 14.3 ··	26.8	Alnilam	276 13.6 S 1 13.1
R 10	47 25.2	293 25.7	31.3	257 18.9	34.7	305 32.2	05.2	259 16.6	26.8	Alphard	218 22.2 S 8 34.0
I 11	62 27.7	308 25.0	30.9	272 20.0	34.2	320 34.0	05.2	274 18.9	26.7			
D 12	77 30.1	323 24.2 N23 30.4		287 21.2 N13 33.7		335 35.9 N23 05.2		289 21.2 N14 26.7		Alphecca	126 33.1 N26 47.4
A 13	92 32.6	338 23.5	30.0	302 22.4	33.2	350 37.8	05.1	304 23.5	26.6	Alpheratz	358 10.9 N28 58.1
Y 14	107 35.0	353 22.8	29.5	317 23.6	32.7	5 39.7	05.1	319 25.8	26.5	Altair	62 33.8 N 8 48.7
15	122 37.5	8 22.0 ··	29.0	332 24.8 ··	32.2	20 41.5 ··	05.0	334 28.1 ··	26.5	Ankaa	353 42.0 S42 25.2
16	137 40.0	23 21.3	28.6	347 25.9	31.7	35 43.4	05.0	349 30.3	26.4	Antares	112 58.4 S26 23.0
17	152 42.4	38 20.6	28.1	2 27.1	31.2	50 45.3	05.0	4 32.6	26.3			
18	167 44.9	53 19.9 N23 27.7		17 28.3 N13 30.7		65 47.2 N23 04.9		19 34.9 N14 26.3		Arcturus	146 19.6 N19 17.8
19	182 47.4	68 19.1	27.2	32 29.5	30.2	80 49.0	04.9	34 37.2	26.2	Atria	108 23.4 S68 59.3
20	197 49.8	83 18.4	26.8	47 30.6	29.7	95 50.9	04.8	49 39.5	26.1	Avior	234 29.2 S59 26.7
21	212 52.3	98 17.7 ··	26.3	62 31.8 ··	29.2	110 52.8 ··	04.8	64 41.8 ··	26.1	Bellatrix	279 00.7 N 6 19.7
22	227 54.8	113 17.0	25.8	77 33.0	28.7	125 54.7	04.7	79 44.1	26.0	Betelgeuse	271 30.3 N 7 24.1
23	242 57.2	128 16.2	25.4	92 34.2	28.2	140 56.6	04.7	94 46.4	25.9			
10 00	257 59.7	143 15.5 N23 24.9		107 35.3 N13 27.7		155 58.4 N23 04.7		109 48.6 N14 25.9		Canopus	264 08.4 S52 41.3
01	273 02.1	158 14.8	24.4	122 36.5	27.2	171 00.3	04.6	124 50.9	25.8	Capella	281 14.0 N45 58.5
02	288 04.6	173 14.1	24.0	137 37.7	26.7	186 02.2	04.6	139 53.2	25.8	Deneb	49 49.2 N45 12.1
03	303 07.1	188 13.3 ··	23.5	152 38.9 ··	26.2	201 04.1 ··	04.5	154 55.5 ··	25.7	Denebola	183 00.6 N14 41.6
04	318 09.5	203 12.6	23.0	167 40.0	25.7	216 05.9	04.5	169 57.8	25.6	Diphda	349 22.6 S18 06.2
05	333 12.0	218 11.9	22.6	182 41.2	25.2	231 07.8	04.5	185 00.1	25.6			
06	348 14.5	233 11.2 N23 22.1		197 42.4 N13 24.7		246 09.7 N23 04.4		200 02.4 N14 25.5		Dubhe	194 24.2 N61 52.3
07	3 16.9	248 10.4	21.6	212 43.6	24.2	261 11.6	04.4	215 04.6	25.4	Elnath	278 46.5 N28 35.3
S 08	18 19.4	263 09.7	21.1	227 44.7	23.7	276 13.4	04.3	230 06.9	25.4	Eltanin	90 57.9 N51 29.6
A 09	33 21.9	278 09.0 ··	20.7	242 45.9 ··	23.1	291 15.3 ··	04.3	245 09.2 ··	25.3	Enif	34 12.9 N 9 46.6
T 10	48 24.3	293 08.3	20.2	257 47.1	22.6	306 17.2	04.3	260 11.5	25.2	Fomalhaut	15 53.1 S29 44.0
U 11	63 26.8	308 07.6	19.7	272 48.3	22.1	321 19.1	04.2	275 13.8	25.2			
R 12	78 29.3	323 06.8 N23 19.2		287 49.4 N13 21.6		336 20.9 N23 04.2		290 16.1 N14 25.1		Gacrux	172 30.1 S56 59.8
D 13	93 31.7	338 06.1	18.8	302 50.6	21.1	351 22.8	04.1	305 18.3	25.0	Gienah	176 19.4 S17 25.5
A 14	108 34.2	353 05.4	18.3	317 51.8	20.6	6 24.7	04.1	320 20.6	25.0	Hadar	149 25.0 S60 16.3
Y 15	123 36.6	8 04.7 ··	17.8	332 53.0 ··	20.1	21 26.6 ··	04.0	335 22.9 ··	24.9	Hamal	328 30.9 N23 21.5
16	138 39.1	23 04.0	17.3	347 54.1	19.6	36 28.4	04.0	350 25.2	24.8	Kaus Aust.	84 18.5 S34 23.6
17	153 41.6	38 03.3	16.8	2 55.3	19.1	51 30.3	04.0	5 27.5	24.8			
18	168 44.0	53 02.5 N23 16.4		17 56.5 N13 18.6		66 32.2 N23 03.9		20 29.8 N14 24.7		Kochab	137 18.1 N74 14.9
19	183 46.5	68 01.8	15.9	32 57.7	18.1	81 34.1	03.9	35 32.1	24.6	Markab	14 04.7 N15 05.3
20	198 49.0	83 01.1	15.4	47 58.8	17.6	96 35.9	03.8	50 34.3	24.6	Menkar	314 43.0 N 4 00.2
21	213 51.4	98 00.4 ··	14.9	63 00.0 ··	17.1	111 37.8 ··	03.8	65 36.6 ··	24.5	Menkent	148 38.5 S36 16.0
22	228 53.9	112 59.7	14.4	78 01.2	16.6	126 39.7	03.8	80 38.9	24.4	Miaplacidus	221 45.6 S69 38.1
23	243 56.4	127 59.0	13.9	93 02.4	16.0	141 41.6	03.7	95 41.2	24.4			
11 00	258 58.8	142 58.2 N23 13.4		108 03.5 N13 15.5		156 43.4 N23 03.7		110 43.5 N14 24.3		Mirfak	309 18.7 N49 46.9
01	274 01.3	157 57.5	12.9	123 04.7	15.0	171 45.3	03.6	125 45.8	24.2	Nunki	76 30.7 S26 19.3
02	289 03.8	172 56.8	12.4	138 05.9	14.5	186 47.2	03.6	140 48.0	24.2	Peacock	54 00.5 S56 48.0
03	304 06.2	187 56.1 ··	11.9	153 07.1 ··	14.0	201 49.1 ··	03.5	155 50.3 ··	24.1	Pollux	244 00.4 N28 04.7
04	319 08.7	202 55.4	11.4	168 08.2	13.5	216 50.9	03.5	170 52.6	24.0	Procyon	245 27.7 N 5 16.7
05	334 11.1	217 54.7	10.9	183 09.4	13.0	231 52.8	03.5	185 54.9	24.0			
06	349 13.6	232 54.0 N23 10.5		198 10.6 N13 12.5		246 54.7 N23 03.4		200 57.2 N14 23.9		Rasalhague	96 30.7 N12 34.6
07	4 16.1	247 53.3	10.0	213 11.8	12.0	261 56.6	03.4	215 59.4	23.8	Regulus	208 11.8 N12 04.3
08	19 18.5	262 52.6	09.5	228 12.9	11.5	276 58.4	03.3	231 01.7	23.8	Rigel	281 37.8 S 8 13.7
S 09	34 21.0	277 51.9 ··	09.0	243 14.1 ··	11.0	292 00.3 ··	03.3	246 04.0 ··	23.7	Rigil Kent.	140 27.3 S60 44.9
U 10	49 23.5	292 51.1	08.4	258 15.3	10.5	307 02.2	03.2	261 06.3	23.7	Sabik	102 42.5 S15 41.8
N 11	64 25.9	307 50.4	07.9	273 16.6	09.9	322 04.1	03.2	276 08.6	23.6			
D 12	79 28.4	322 49.7 N23 07.4		288 17.6 N13 09.4		337 05.9 N23 03.2		291 10.9 N14 23.5		Schedar	350 10.8 N56 24.9
A 13	94 30.9	337 49.0	06.9	303 18.8	08.9	352 07.8	03.1	306 13.1	23.5	Shaula	96 57.4 S37 05.2
Y 14	109 33.3	352 48.3	06.4	318 20.0	08.4	7 09.7	03.1	321 15.4	23.4	Sirius	258 57.4 S16 41.4
15	124 35.8	7 47.6 ··	05.9	333 21.1 ··	07.9	22 11.6 ··	03.0	336 17.7 ··	23.3	Spica	158 59.0 S11 03.0
16	139 38.2	22 46.9	05.4	348 22.3	07.4	37 13.4	03.0	351 20.0	23.3	Suhail	223 12.1 S43 21.0
17	154 40.7	37 46.2	04.9	3 23.5	06.9	52 15.3	02.9	6 22.3	23.2			
18	169 43.2	52 45.5 N23 04.4		18 24.6 N13 06.4		67 17.2 N23 02.9		21 24.5 N14 23.1		Vega	80 56.5 N38 45.8
19	184 45.6	67 44.8	03.9	33 25.8	05.9	82 19.0	02.9	36 26.8	23.1	Zuben'ubi	137 34.5 S15 57.2
20	199 48.1	82 44.1	03.4	48 27.0	05.3	97 20.9	02.8	51 29.1	23.0			
21	214 50.6	97 43.4 ··	02.9	63 28.2 ··	04.8	112 22.8 ··	02.8	66 31.4 ··	22.9		S.H.A.	Mer. Pass.
22	229 53.0	112 42.7	02.3	78 29.3	04.3	127 24.7	02.7	81 33.7	22.8	Venus	245 15.8 14 28
23	244 55.5	127 42.0	01.8	93 30.5	03.8	142 26.5	02.7	96 35.9	22.8	Mars	209 35.7 16 48
										Jupiter	257 58.7 13 34
Mer. Pass.	6 46.9	v −0.7 d 0.5		v 1.2 d 0.5		v 1.9 d 0.0		v 2.3 d 0.1		Saturn	211 49.0 16 38

1978 JUNE 9, 10, 11 (FRI., SAT., SUN.)

G.M.T.	ARIES G.H.A.	VENUS −3.4 G.H.A.	Dec.	MARS +1.5 G.H.A.	Dec.	.
d h	° ′	° ′	° ′	° ′	° ′	
9 00	257 00.5	143 33.0 N23	35.7	107 07.1 N13	39.8	1!
01	272 03.0	158 32.3	35.3	122 08.3	39.3	17
02	287 05.5	173 31.6	34.9	137 09.4	38.8	1′
03	302 07.9	188 30.8 ··	34.4	152 10.6 ··	38.3	
04	317		·· °	167 11.8		8.
		83 01.1			.0	96
21	213 51.4	98 00.4 ··	14.9	63 00.0 ··	17.1	111
22	228 53.9	112 59.7	14.4	78 01.2	16.6	126
23	243 56.4	127 59.0	13.9	93 02.4	16.0	141
11 00	258 58.8	142 58.2 N23	13.4	108 03.5 N13	15.5	156
01	274 01.3	157 57.5	12.9	123 04.7	15.0	17
02	289 03.8	172 56.8	12.4	138 05.9	14.5	18
03	304 06.2	187 56.1 ··	11.9	153 07.1 ··	14.0	2(
04	319 08.7	202 55.4	11.4	168 08.2	13.5	2
05	334 11.1	217 54.7	10.9	183 09.4	13.0	2
06	349 13.6	232 54.0 N23	10.5	198 10.6 N13	12.5	2
07	4 16.1	247 53.3	10.0	213 11.8	12.0	2
08	19 18.5	262 52.6	09.5	228 12.9	11.5	2
S 09	34 21.0	277 51.9 ··	09.0	243 14.1 ··	11.0	2!
U 10	49 23.5	292 51.1	08.4	258 15.3	10.5	30
N 11	64 25.9	307 50.4	07.9	273 16.4	09.9	32
D 12	79 28.4	322 49.7 N23	07.4	288 17.6 N13	09.4	33;
A 13	94 30.9	337 49.0	06.9	303 18.8	08.9	352
Y 14	109 33.3	352 48.3	06.4	318 20.0	08.4	7
15	124 35.8	7 47.6 ··	05.9	333 21.1 ··	07.9	22
16	139 38.2	22 46.9	05.4	348 22.3	07.4	37
17	154 40.7	37 46.2	04.9	3 23.5	06.9	5;
18	169 43.2	52 45.5 N23	04.4	18 24.6 N13	06.4	6
19	184 45.6	67 44.8	03.9	33 25.8	05.9	8
20	199 48.1	82 44.1	03.4	48 27.0	05.3	!
21	214 50.6	97 43.4 ··	02.9	63 28.2 ··	04.8	1:
22	229 53.0	112 42.7	02.3	78 29.3	04.3	1:
23	244 55.5	127 42.0	01.8	93 30.5	03.8	1
Mer. Pass.	h m 6 46.9	v −0.7	d 0.5	v 1.2	d 0.5	

Figure 8–3. Excerpt from Nautical Almanac showing Astronomical Data for the Planet Venus on June 11 at 23^h G.M.T.

0.1 minutes must be applied to the *negative* 1.7' apparent altitude correction to arrive at the final correction, and the correct Ho.

The GHA and Dec information is taken from the left hand side of the almanac's daily pages, a specimen of which is shown in Figure 8-2. An excerpt of this table, applicable to our practical example is shown in Figure 8-3.

You will see that the GHA and Dec data is presented for each of the navigational planets by name, and is tabulated like the sun. At the bottom of each column is a single *v* and *d* value which apply for the three day period. You will readily find the GHA of 127-42.0 and the Dec of 23-01.8 N., for our example, opposite the hours of GMT (23ʰ). The increment for minutes and seconds of GMT is found in the yellow section at the back of the almanac (Figure 8-4) where you will notice the planets share the same column with the sun.

For the 56ᵐ 14ˢ of GMT, the GHA increment is found to be 14-03.5. As in the exercise with the moon, the *v* and *d* corrections are determined by entering the right hand columns of the table with the appropriate values from the daily page (Figure 8-3), and are applied to the GHA and Dec, respectively, in accordance with their proper signs which should be observed carefully.

The remainder of the procedure for reducing your planet sight follows the same routine you learned for the sun and moon. Applying your Assumed Longitude of 72-44.8 W to your calculated GHA of 141-44.8, you arrive at an LHA of 69, which, with your calculated Dec of 23-01.3 N, and your Assumed Latitude of 41° N, gives you the figures to enter the proper page of the sight reduction table (Figure 8-5).

56 m	SUN PLANETS	ARIES	MOON	v or Corrⁿ d		v or Corrⁿ d		v or Corrⁿ d	
s	° ′	° ′	° ′	′ ′		′ ′		′ ′	
00	14 00·0	14 02·3	13 21·7	0·0	0·0	6·0	5·7	12·0	11·3
01	14 00·3	14 02·6	13 22·0	0·1	0·1	6·1	5·7	12·1	11·4
02	14 00·5	14 02·8	13 22·2	0·2	0·2	6·2	5·8	12·2	11·5
03	14 00·8	14 03·1	13 22·4	0·3	0·3	6·3	5·9	12·3	11·6
04	14 01·0	14 03·3	13 22·7	0·4	0·4	6·4	6·0	12·4	11·7
05	14 01·3	14 03·6	13 22·9	0·5	0·5	6·5	6·1	12·5	11·8
06	14 01·5	14 03·8	13 23·2	0·6	0·6	6·6	6·2	12·6	11·9
07	14 01·8	14 04·1	13 23·4	0·7	0·7	6·7	6·3	12·7	12·0
08	14 02·0	14 04·3	13 23·6	0·8	0·8	6·8	6·4	12·8	12·1
09	14 02·3	14 04·6	13 23·9	0·9	0·8	6·9	6·5	12·9	12·1
10	14 02·5	14 04·8	13 24·1	1·0	0·9	7·0	6·6	13·0	12·2
11	14 02·8	14 05·1	13 24·4	1·1	1·0	7·1	6·7	13·1	12·3
12	14 03·0	14 05·3	13 24·6	1·2	1·1	7·2	6·8	13·2	12·4
13	14 03·3	14 05·6	13 24·8	1·3	1·2	7·3	6·9	13·3	12·5
14	14 03·5	14 05·8	13 25·1	1·4	1·3	7·4	7·0	13·4	12·6
15	14 03·8	14 06·1	13 25·3	1·5	1·4	7·5	7·1	13·5	12·7
16	14 04·0	14 06·3	13 25·6	1·6	1·5	7·6	7·2	13·6	12·8
17	14 04·3	14 06·6	13 25·8	1·7	1·6	7·7	7·3	13·7	12·9
18	14 04·5	14 06·8	13 26·0	1·8	1·7	7·8	7·3	13·8	13·0
19	14 04·8	14 07·1	13 26·3	1·9	1·8	7·9	7·4	13·9	13·1

Figure 8–4. Excerpt from Nautical Almanac Increments and Corrections Tables showing Planet's increment for 56ᵐ 14ˢ and v and d corrections

The tabular Hc of 30-21 is corrected for the incremental minutes of declination (1.3) at the altitude differential, *d,* of +35. The correction of 1′, found in Table 5 (Figure 8-6), produces an Hc of 30-22.0, which, when compared with your previously determined Ho of 30-16.1, yields an intercept of 5.9 "Away".

In our practical example, you will recall that the LHA was 69, which is *less* than 180°, so that Zn, in accordance with the rules in the upper corner of the sight reduction

Figure 8–5. Excerpt from Table of Computed Altitudes and Azimuths, Pub. No. 249, Vol. III, for Latitude 41°, Declination 23°—Same Name as Latitude—and Local Hour Angle, 69

N. Lat. {LHA greater than 180°......... Zn=Z / LHA less than 180°...........Zn=360−Z}

LAT 41°

DECLINATION (15°-29°) SAME NAME AS LATITUDE

LHA	21° Hc	d	Z	22° Hc	d	Z	23° Hc	d	Z	24° Hc	d	Z	25° Hc	d	Z	LHA
	° ′	′	°	° ′	′	°	° ′	′	°	° ′	′	°	° ′	′	°	
0	70 00	+60	180	71 00	+60	180	72 00	+60	180	73 00	+60	180	74 00	+60	180	360
1	69 59	60	177	70 59	60	177	71 59	60	177	72 59	60	177	73 59	60	177	359
2	69 56	60	175	70 56	59	174	71 55	60	174	72 55	60	174	73 55	60	173	358
3	69 50	60	172	70 50	60	172	71 49	59	171	72 49	59	171	73 48	60	170	357
4	69 43	59	169	70 42	59	169	71 41	59	168	72 40	59	168	73 39	59	167	356
5	69 33	+59	167	70 32	+59	166	71 31	+59	165	72 30	+58	165	73 28	+58	164	355
6	69 22			⁻⁻ ²⁰	58	163	71 18	59	163	72 17						354
							71 04	57								
58	37 29	35	94	38 04	35	93	38 49	34	92	39 13	33	91	40 32	33	90	302
59	36 44	35	93	37 19	34	92	37 53	35	91	38 28	34	90	39 01	33	89	301
60	35 58	+36	93	36 34	+34	92	37 08	+34	91	37 42	+34	90	38 16	+33	88	300
61	35 13	35	92	35 48	35	91	36 23	34	90	36 57	34	89	37 31	33	88	299
62	34 28	35	91	35 03	35	90	35 38	34	89	36 12	33	88	36 45	33	87	298
63	33 43	35	91	34 18	34	90	34 52	34	89	35 26	34	88	36 00	33	87	297
64	32 57	35	90	33 32	35	89	34 07	34	88	34 41	34	87	35 15	33	86	296
65	32 12	+35	89	32 47	+35	88	33 22	+34	87	33 56	+34	86	34 30	+33	85	295
66	31 27	35	89	32 02	35	88	32 37	34	87	33 11	34	86	33 45	33	85	294
67	30 42	35	88	31 17	34	87	31 51	35	86	32 26	34	85	33 00	33	84	293
68	29 56	35	87	30 31	35	86	31 06	35	85	31 41	34	84	32 15	33	84	292
69	29 11	35	87	29 46	35	86	30 21	35	85	30 56	34	84	31 30	33	83	291
	21°			22°			23°			24°			25°			

NAVIGATOR'S WORKBOOK

DATE	June 11
BODY	Venus
hs	30 - 22.0
IC	- 1.4
D	- 2.9
ha	30 - 17.7
R	- 1.7
add'l corr	+0.1
Ho	30 - 16.1
W	23 - 56 -14
corr	00
GMT	23 - 56 - 14
gha [v]	127 - 42.0 +0.7
incr	14 - 03.5
v corr	- 0.7
GHA	141 - 44.8
aλ	72 - 44.8
LHA	69
dec [d]	23 - 01.8 -0.5
d corr	- 0.5
Dec	23 - 01.3 N
aL	41 N
Tab Hc	30 - 21
corr	+ 01
Hc	30 - 22.0
Ho	30 - 16.1
a	5.9 A
Zn	275

Workform for Planet Observation

TABLE 5.—Correction to Tabulated Altitude

32 33	34 35 36	37 38 39	40 41 42	43		54	55 56 57	58 59 60	$\frac{d}{'}$
0 0	0 0 0	0 0 0	0 0 0	0		0	0 0 0	0 0 0	0
1 1	1 1 1	1 1 1	1 1 1	1		1	1 1 1	1 1 1	1
1 1	1 1 1	1 1 1	1 1 1	1		2	2 2 2	2 2 2	2
2 2	2 2 2	2 2 2	2 2 2	2		3	3 3 3	3 3 3	3
2 2	2 2 2	2 3 3	3 3 3	3		4	4 4 4	4 4 4	4
3 3	3 3 3	3 3 3	3 3 4	4		4	5 5 5	5 5 5	5
3 3	3 4 4	4 4 4	4 4 4	4		5	6 6 6	6 6 6	6
4 4	4 4 4	4 4 5	5 5 5	5		6	6 7 7	7 7 7	7
4 4	5 5 5	5 5 5	5 5 6	6		7	7 7 8	8 8 8	8
5 5	5 5 5	6 6 6	6 6 6	6		8	8 8 9	9 9 9	9
5 6	6 6 6	6 6 6	7 7 7	7		9	9 9 10	10 10 10	10
6 6	6 6 7	7 7 7	7 8 8	8		0	10 10 10	11 11 11	11
6 7	7 7 7	7 8 8	8 8 8	9		1	11 11 11	12 12 12	12
7 7	7 8 8	8 8 8	9 9 9	9 10		2	12 12 12	13 13 13	13
8	8 8 8	9 9 9	9 10 10	10 10			13 13 13	14 14 14	14
8	8 9		10 10 10	11 11				15	15

Figure 8–6. Excerpt from Table 5, Pub. No. 249, showing Correction for Tabulated Altitude for d of 35, and Incremental Minutes, 1

table (Figure 8-5), equals 360-Z, for a true azimuth of 275°. Plotting of the 5.9′ intercept from the AP—in this case *away* from the direction of the true azimuth—and construction of the line of position at right angles to the azimuth at that point, proceeds exactly as you have done in your earlier sights.

9. The Stars

It has always been something of a wonder to me why most people, when they think of celestial navigation, think first in terms of the stars. Star sights do offer the attractive feature that several can be observed almost simultaneously, producing lines of position which can be crossed for an instantaneous fix. But, except for that attraction, and the fact that with Vol. I of Pub. No. 249 selected star sights are particularly easy to solve, my experience has been that shooting the stars is a lot more demanding than the sun or moon for equivalent results.

Let's talk about the easy part first—solving the star sight for its line of position—and follow the steps with a practical example. There are two options from which to choose when you get to the sight reduction tables, so we'll work our sight with both, commenting as we go. The workforms illustrated can be followed in parallel until the procedure starts to diverge at the sight reduction stage and this will present a good comparison of the alternatives.

At evening twilight on June 9th, 1978, in DR position Latitude 40° 55′ North, Longitude 70° 21′ West, you observe the star Arcturus. The GMT is 00-56-34 on June 10th (remember you are four hours slow of Greenwich Mean Time at that position during the time of year when Daylight Saving Time prevails), the sextant altitude is 66-22.9, its Index Correction is −1.4 and your height of eye is nine feet.

The sextant altitude is corrected for IC and Dip in exactly the same way as for the other bodies, and the resulting ha is entered into the STARS AND PLANETS column of the Altitude Correction Table, found inside the front cover of the almanac, to obtain the R correction. Figure 9-1 shows a portion of the appropriate table from which you will take your D correction of −2.9′ and your R of −0.4′.

You will notice that there is just a single R correction to correct the ha of a star to its Ho, and that it is always negative. Thus, for our exercise, the Ho of Arcturus becomes 66-18.2.

Finding the GHA of your star is slightly different from your previous entries into the almanac's daily pages. Since the stars change position with respect to each other only at a very slow rate, the almanac designers, in order to streamline that splendid publication, have chosen to present the astronomical data for an arbitrary, single point on the celestial sphere, and relate all the stars' positions to it. The point, which is called the "First Point of Aries" or "Vernal Equinox", abbreviated ♈, is

ALTITUDE CORRECTION TABLES 10°–90°

STARS AND PLANETS			DIP				
App. Alt.	Corrⁿ	App. Alt.	Additional Corrⁿ	Ht. of Eye	Corrⁿ	Ht. of Eye	Ht. of Eye	Corrⁿ

App. Alt.	Corrⁿ	App. Alt. Additional Corrⁿ	Ht. of Eye / Corrⁿ	Ht. of Eye	Ht. of Eye / Corrⁿ
° ′		**1978**	m	ft.	m ′
9 56	−5·3	**VENUS**	2·4 −2·8	8·0	1·0 − 1·8
10 08	−5·2	Jan. 1-July 20	2·6 −2·9	8·6	1·5 − 2·2
10 20	−5·1	° ′	2·8 −3·0	9·2	2·0 − 2·5
10 33	−5·0	42 + 0·1	3·0 −3·1	9·8	2·5 − 2·8
10 46			3·2 −2·2	10·5	2·0 − 3·0
	−0·6			60·5	
00 28	−0·5		18 4 −7·6	62·1	130 −11·1
65 08	−0·4		18·8 −7·7	63·8	135 −11·3
70 11	−0·3		19·3 −7·8	65·4	140 −11·5
75 34	−0·2		19·8 −7·9	67·1	145 −11·7
81 13	−0·1		20·4 −8·0	68·8	150 −11·9
87 03	0·0		20·9 −8·1	70·5	155 −12·1
90 00			21·4		

App. Alt.   Apparent altitude

Sextant altitude corrected for index error and dip.

Figure 9–1. Portion of Altitude Correction Tables from Nautical Almanac showing Dip correction (−2.9′ for Ht. of Eye of 9 ft.) and Star's Refraction correction (−0.4′ for App. Alt. of 66° 18.6′)

tabulated as if it were a celestial body. All the stars are positioned from that point by their Sidereal Hour Angle (SHA)—thus the simple formula:

GHA Aries + SHA Star = GHA Star

In Figure 8-2 the specimen page from the almanac showed the presentation of the astronomical data for the stars and planets. The excerpt in Figure 9-2 of a portion of that page shows the information needed for the solution of our practical example.

From the ARIES column, for the GMT hours of your star sight, you extract the GHA of Aries. In our example the GHA would be 257-59.7 at 00 hours on the GMT date of June 10. The incremental minutes and seconds are dealt with in the ARIES column of the yellow Increments and Corrections tables in the back of the almanac. A glance at Figure 9-3, a table with which you are already familiar, will show for 56 minutes and 34 seconds, an increment for Aries of 14-10.8.

Having found the GHA♈, you now have your two options: the first and simplest is to use Vol. I of Pub. No. 249, providing you have selected one of the seven stars listed for the degree of LHA. It is highly likely that you will find your star in the table as you probably will have chosen it in advance for that very reason. On the other hand, if the declination of the star is less than 30°, North or South, you may elect to work the sight by the method of Vols. II and III with which you have worked all the bodies of the solar system.

To put your choice in perspective, consider that in the *Nautical Almanac*'s daily pages there are listed the positions of 57 navigational stars—the "Selected Stars"— which have been chosen for their relative brightness and for their distribution throughout the sky. You will seldom, if ever, use all of these. Of that 57, seven stars are selected in Vol. I (Pub. No. 249) for each degree of LHA, using a total of 41 altogether. Thirty stars, of the "selected" 57, including seven lesser ones which do not

1978 JUNE 9, 10, 11 (FRI., SAT., SUN.)

G.M.T.	ARIES G.H.A.	VENUS G.H.A.	+0.8 Dec.	STARS Name	S.H.A.	Dec.
d h	° ′	° ′	° ′		° ′	° ′
9 00	257 00.5	143 33.0 N	114 27.4	Acamar	315 38.8	S40 23.4
01	272 03.0	158 32.3	27.4	Achernar	335 46.9	S57 20.6
02	287 05.5	173 31.6	27.3	Acrux	173 38.6	S62 59.1
03	302 07.9	188 30.8 ·	· 27.2	Adhara	255 33.7	S28 56.8
04	317 10.4	203 30.1	27.2	Aldebaran	291 20.1	N16 27.8
05	332 12.9	218 29.4	27.1			
06	347 15.3	233 28.6 N23	1 27.0	Alioth	166 43.7	N56 04.9
07	2 17.8	248 27.9	27.0	Alkaid	153 19.5	N49 25.5
08	17 20.3	263 27.2	26.9	Al Na'ir	28 16.8	S47 03.7
F 09	32 22.7	278 26.4 ··	26.8	Alnilam	276 13.6	S 1 13.1
R 10	47 25.2	293 25.7	26.8	Alphard	218 22.2	S 8 34.0
I 11	62 27.7	308 25.0	26.7			
D 12	77 30.1	323 24.2 N23	26.7	Alphecca	126 33.1	N26 47.4
A 13	92 32.6	338 23.5	26.6	Alpheratz	358 10.9	N28 58.1
Y 14	107 35.0	353 22.8	26.5	Altair	62 33.8	N 8 48.7
15	122 37.5	8 22.0 ··	26.5	Ankaa	353 42.0	S42 25.2
16	137 40.0	23 21.3	26.4	Antares	112 58.4	S26 23.0
17	152 42.4	38 20.6	26.3			
18	167 44.9	53 19.9 N2	4 26.3	Arcturus	146 19.6	N19 17.8
19	182 47.4	68 19.1	26.2	Atria	108 23.4	S68 59.3
20	197 49.8	83 18.4	26.1	Avior	234 29.2	S59 26.7
21	212 52.3	98 17.7 ··	26.1	Bellatrix	279 00.7	N 6 19.7
22	227 54.8	113 17.0	26.0	Betelgeuse	271 30.3	N 7 24.1
23	242 57.2	128 16.2	25.9			
10 00	257 59.7	143 15.5 N23	25.9	Canopus	264 08.4	S52 41.3
01	273 02.1	158 14.8	25.8	Capella	281 14.0	N45 58.5
02	288 04.6	173 14.1	25.8	Deneb	49 49.2	N45 12.1
03	303 07.1	188 13.3 ··	25.7	Denebola	183 00.6	N14 41.6
04	318 09.5	203 12.6	25.6	Diphda	349 22.6	S18 06.2
05	333 12.0	218 11.9	25.6			
06	348 14.5	233 11.2 N2	25.5	Dubhe	194 24.2	N61 52.3
07	3 16.9	248 10.4	25.4	Elnath	278 46.5	N28 35.3
S 08	18 19.4	263 09.7	25.4	Eltanin	90 57.9	N51 29.6
A 09	33 21.9	278 09.0 ·	· 25.3	Enif	34 12.9	N 9 46.6
			25.2		73 3	S29 44.0

Figure 9–2. Excerpt from Nautical Almanac showing Astronomical Data for the star Arcturus on June 10 at 00h G.M.T.

56ᵐ SUN PLANETS	ARIES	MOON	v or Corrⁿ d	v or Corrⁿ d	v or Corrⁿ d
s ° ′	° ′	° ′	′ ′	′ ′	′ ′
00 14 00·0	14 02·3	13 21·7	0·0 0·0	6·0 5·7	12·0 11·3
01 14 00·3	14 02·6	13 22·0	0·1 0·1	6·1 5·7	12·1 11·4
02 14 00·5	14 02·8	13 22·2	0·2 0·2	6·2 5·8	12·2 11·5
03 14 00·8	14 03·1	13 22·4	0·3 0·3	6·3 5·9	12·3 11·6
04					
	14 07·0				
31 14 07·8	14 10·1	13 29·1	3·1 2·9	9·1 8·6	15·1 14·2
32 14 08·0	14 10·3	13 29·4	3·2 3·0	9·2 8·7	15·2 14·3
33 14 08·3	14 10·6	13 29·6	3·3 3·1	9·3 8·8	15·3 14·4
34 14 08·5	14 10·8	13 29·8	3·4 3·2	9·4 8·9	15·4 14·5
35 14 08·8	14 11·1	13 30·1	3·5 3·3	9·5 8·9	15·5 14·6
36 14 09·0	14 11·3	13 30·3	3·6 3·4	9·6 9·0	15·6 14·7
37 14 09·3	14 11·6	13 30·6	3·7 3·5	9·7 9·1	15·7 14·8
38 14 09·5	14 11·8	13 30·8	3·8 3·6	9·8 9·2	15·8 14·9
39 14 09·8		13 31·0	3·9 3·7		15·9 15·0

Figure 9–3. Excerpt from Nautical Almanac Increments and Corrections Tables showing Aries increment for 56ᵐ 34ˢ

appear in Vol. I, can be worked by Vols. II or III. Thus, only nine, and all those minor, of the "selected" list can't be worked with one or the other volumes of Pub. No. 249. Never, in my experience, has this been a problem, nor would it be a consideration in the choice of Pub. No. 249 against all the other sight reduction tables. With a little pre-planning you will almost always use Vol. I for your star sights, but since, of the 19 first magnitude (brightest) stars in the almanac list, an even dozen can be worked with either volume, I believe you will agree that it is worthwhile to understand the method for each.

In the Vol. I method, you apply your Assumed Longitude to the GHA of Aries, choosing a longitude, as usual, to make the LHA a whole degree. This result is identified as the Local Hour Angle of Aries (LHA ♈) and is all you need besides your Assumed Latitude (41°N), and the name of your star, Arcturus, to enter the table. With our GHA ♈ of 272-10.5, we assumed a longitude of 70° 10.5' W, giving us an LHA ♈ of 202.

Figure 9-4 shows a portion of the page from Vol. I covering Latitude 41° N., your aL. Finding your LHA ♈ (202) in the column at the left of the tabulation, you take out the Hc and Zn opposite that number, directly under the name of the star you have observed. That's all there is to it!

While there are no further corrections or interpolation needed with the figures from Vol. I, you should notice that your issue will be for a certain "epoch" year, the volume being updated every five years. If you are more than a year either side of the epoch year as listed on the cover, you may wish to correct your *fix* as shown in Table 5 in the back of Vol. I for "Precession and Nutation", two wonderful words which simply mean that while the stars' positions can be considered fixed for short periods, they do wander slowly and over a period of years this ought to be taken into account. In all candor, I have to tell you that a great number of small-boat navigators simply ignore this correction although it can admit errors of two or three miles in the distant years. In any case, you should remember that this correction applies to the *fix,* not the line of position, and

⟍LAT 41°N

LHA ♈	Hc Zn	Hc Zn	Hc Zn	Hc Zn	Hc Zn	Hc Zn	Hc Zn
	Kochab	♦VEGA	ARCTURUS	♦SPICA	REGULUS	♦POLLUX	CAPELLA
180	51 27 017	18 33 054	54 02 117	34 26 155	51 58 229	36 56 277	21 43 313
181	51 40 017	19 10 055	54 43 118	34 45 156	51 24 230	36 11 278	21 10 313
182	51 54 017	19 47 055	55 22 119	35 03 157	50 49 231	35 26 278	20 37 314
183	52 06 016	20 24 056	56 02 120	35 21 158	50 14 232	34 42 279	20 04 314
184	52 19 016	21 02 056	56 40 122	35 37 159	49 38 234	33 57 280	19 32 315
185	52 32 016	21 39 057	57 19 123	35 53 160	49 01 235	33 12 280	19 00 315
186	52 44 016	22 17 057	57 56 124	36 07 162	48 24 236	32 28 281	18 28 316
187	52 56 015	22 56 058	58 33 126	36 21 163	47 46 237	31 43 281	17 56 316
188	53 08 015	23 34 058	59 10 127	36 34 164	47 08 238	30 59 282	17 25 317
189	53 19 015	24 13 059	59 46 128	36 46 165	46 29 239	30 15 282	16 54 317
190	53 30 014	24 51 059	60 21 130	36 57 166	45 50 240	29 31 283	16 23 318
191	53 42 014	25 30 060	60 55 131	37 07 168	45 11 241	28 47 284	15 53 318
192	53 52 014	26 10 060	61 29 133	37 17 169	44 31 242	28 03 284	15 23 319
193	54 03 013	26 49 061	62 01 135	37 25 170	43 51 243	27 19 285	14 53 319
194	54 13 013	27 29 061	62 33 136	37 32 171	43 10 244	26 35 285	14 23 320
	♦VEGA	Rasalhague	ARCTURUS	♦SPICA	REGULUS	♦POLLUX	Dubhe
195	28 08 062	24 23 094	63 04 138	37 39 173	42 29 245	25 51 286	62 43 330
196	28 48 062	25 08 095	63 33 140	37 44 174	41 48 246	25 08 286	62 20 329
197	29 28 063	25 53 096	64 02 142	37 49 175	41 06 247	24 24 287	61 56 329
198	30 09 063	26 38 097	64 29 144	37 52 176	40 24 248	23 41 287	61 32 328
199	30 49 063	27 23 097	64 56 146	37 55 177	39 42 249	22 58 288	61 08 327
200	31 30 064	28 08 098	65 21 148	37 56 179	39 00 250	22 15 289	60 44 327
201	32 10 064	28 53 099	65 44 150	37 57 180	38 17 251	21 32 289	60 19 327
202	32 51 065	29 37 099	66 06 152	37 56 181	37 35 251	20 49 290	59 54 326
203	33 32 065	30 22 100	66 27 154	37 55 182	36 52 252	20 07 290	59 28 326
204	34 13 066	31 07 101	66 46 156	37 52 184	36 08 253	19 24 291	59 03 326
205	34 55 066	31 51 102	67 03 159	37 49 185	35 25 254	18 42 291	58 37 325
206	35 36 067	32 35 102	67 19 161	37 44 186	34 41 255	18 00 292	58 11 325
207	36 18 067	33 19 103	67 33 163	37 39 187	33 58 255	17 18 292	57 44 324
208	37 00 067	34 03 104	67 45 166	37 33 189	33 14 256	16 36 293	57 18 324
209	37 41 068	34 47 105	67 55 168	37 26 190	32 30 ⸺	15 55 294	56 51 32⸱

Figure 9–4. Excerpt of Table of Computed Altitudes and Azimuths, Pub. No. 249, Vol. I, Epoch 1980.0, for Latitude 41° N, Local Hour Angle of Aries 202, and the star Arcturus

only to fixes deduced from Vol. I, not the other volumes.

Concluding the first part of our exercise, you have compared the Hc from Vol. I (66-06.0) with the previously determined Ho (66-18.2), to arrive at the intercept

NAVIGATOR'S WORKBOOK

DATE	June 10
BODY	Arcturus
hs	66 - 22.9
IC	- 1.4
D	-2.9
ha	66 - 18.6
R	-0.4
Ho	66 - 18.2
W	00 - 56 -34
corr	00
GMT	00 - 56 - 34
ghaϒ	257 - 59.7
incr	14 - 10.8
GHAϒ	272 - 10.5
SHA ★	146 - 19.6
	(-360)
GHA ★	58 - 30.1
	(+360)
	418 - 30.1
aλ	70 - 30.1
LHA ★	348
Dec	19 - 17.8 N
aL	41 N
Tab Hc	65 - 44
corr	16
Hc	66 - 00.0
Ho	66 - 18.2
a	18.2 T
Zn	151°

DATE	June 10
BODY	Arcturus
hs	66 - 22.9
IC	- 1.4
D	-2.9
ha	66 - 18.6
R	-0.4
Ho	66 - 18.2
W	00 - 56 - 34
corr	00
GMT	00 - 56 - 34
ghaϒ	257 - 59.7
incr	14 - 10.8
GHAϒ	272 - 10.5
aλ	70 - 10.5
LHAϒ	202
aL	41 N
Hc	66 - 060
Ho	66 - 18.2
a	12.2 T
Zn	152°

*Workform for Star
by Vol. I, Pub. No. 249*

*Workform for Star
by Vol. II/III, Pub. No. 249*

of 12.2′ "Toward" the True Azimuth (Zn) of 152°. The plot, as before, starts from your AP, is stepped off the length of the intercept and the line of position constructed at right angles at that point.

In the procedure for Vols. II or III, after you had extracted the GHA♈ from the almanac (Figure 9-2), and corrected it for the incremental minutes and seconds of GMT (Figure 9-3), you add the SHA of the star as shown in the star list on the daily page (Figure 9-2) and arrive at the GHA of the star. Next, you apply your Assumed Longitude to that GHA, and, in the usual manner, obtain the LHA of the star. As you compare the star workforms, you will notice that to arrive at a whole degree of LHA, a somewhat different Assumed Longitude from your Vol. I solution was used. You also end up with different LHA's and that is because in one case you derive the LHA of Aries, while in the other you derive the LHA of the star itself. It all comes out in the wash, however, and the position lines will be virtually coincident as long as you remember to step off the proper intercept from the *corresponding* assumed position.

You will notice in the Vol. II/III workform on page 76 that in combining SHA★ with GHA♈ to obtain GHA★, 360° was subtracted from the total to arrive at the correct GHA★. Because the resulting GHA★ was smaller that the aλ which is west and subtractive, 360° had to be added. Since, by inspection, the navigator might have recognized that by omitting the subtraction and later re-addition of 360°, the result would have been the same LHA★, he might have skipped the extra steps which I included for clarity. After obtaining the LHA of your star, the declination is taken directly from the star list (Figure 9-2) after the name of the star and, assuming a latitude nearest your DR position (say, 41° N.), you have the three figures you need to enter Vol. III in the regular way. Figure 9-5 shows a portion of the appropriate page from Vol. III in which you can find the data to complete your star sight reduction.

At Latitude 41°, with the declination of the same name

Figure 9–5. Excerpt from Table of Computed Altitudes and Azimuths, Pub. No. 249, Vol. III, for Latitude 41°, Declination 19°—Same Name as Latitude—and Local Hour Angle, 348

N. Lat. {LHA greater than 180°:........Zn=Z} {LHA less than 180°:............Zn=360-Z}   LAT 41°

DECLINATION (15°-29°) SAME NAME AS LATITUDE

| | 15° | | | 16° | | | 17° | | | 18° | | | 19° | | | 20° | | |
LHA	Hc	d	Z	Hc	d	Z	Hc	d	Z	Hc	d	Z	Hc	d	Z	Hc	d	Z	LHA
	° ′	′	°	° ′	′	°	° ′	′	°	° ′	′	°	° ′	′	°	° ′	′	°	
0	64 00	+60	180	65 00	+60	180	66 00	+60	180	67 00	+60	180	68 00	+60	180	69 00	+60	180	360
1	63 59	60	178	64 59	60	178	65 59	60	178	66 59	60	178	67 59	60	178	68 59	60	177	359
2	63 57	59	176	64 56	60	176	65 56	60	175	66 56	60	175	67 56	60	175	68 56	60	175	358
3	63 52	60	173	64 52	60	173	65 52	59	173	66 51	60	173	67 51	60	173	68 51	59	172	357
4	63 46	60	171	64 46	60	171	65 45	60	171	66 45	59	170	67 44	60	170	68 44	59	170	356
5	63 38	+60	169	64 38	+59	169	65 37	+59	168	66 36	+59	168	67 35	+59	168	68 34	+59	167	355
6	63 29	59	167	64 28	59	167	65 27	59	166	66 26	59	166	67 25	58	165	68 23	59	165	354
7	63 18	59	165	64 17	58	164	65 15	59	164	66 14	58	163	67 12	58	163	68 10	58	162	353
8	63 05	59	163	64 04	58	162	65 02	58	162	66 00	58	161	66 58	57	160	67 55	58	160	352
9	62 51	58	161	63 49	58	160	64 47	57	159	65 44	58	159	66 42	57	158	67 39	57	157	351
10	62 35	+58	159	63 33	+57	158	64 30	+57	157	65 27	+57	157	66 24	+56	156	67 20	+57	155	350
11	62 18	57	157	63 15	57	156	64 12	56	155	65 08	57	154	66 05	55	154	67 00	56	153	349
12	61 59	57	155	62 56	56	154	63 52	56	153	64 48	56	152	65 44	55	151	66 39	55	151	348
13	61 39	56	153	62 35	56	152	63 31	55	151	64 26	55	150	65 21	55	149	66 16	54	148	347
14	61 18	55	151	62 13	55	150	63 08	55	149	64 03	54	148	64 57	54	147	65 51	54	146	346
15	60 55	+55	149	61 50	+55	148	62 45	+54	147	63 39	+53	146	64 32	+54	145	65 26	+52	144	345
16	60 31	55	147	61 26	53	146	62 19	54	145	63 13	53	144	64 06	52	143	64 58	52	142	344
17	60 06	54	146	61 00	53	145	61 53	53	144	62 46	52	143	63 38	52	142	64 30	51	140	343
18	59 40	53	144	60 33	53	143	61 26	52	142	62 18	52	141	63 10	51	140	64 01	50	139	342
19			142	60 05	52	141	60 57			61 49	51	139	62 40			50	137	341

(North), and Arcturus' Dec of 19-17.8, the Tab Hc will be seen to be 65-44, the altitude difference, *d,* +55, and the uncorrected azimuth (Z), 151°. In Table 5, shown in Figure 9-6, the correction for the 17.8 incremental minutes of declination is seen to be 16'. This yields a final Hc of exactly 66-00 which, when compared with the Ho of 66-18.2 you determined earlier, gives you an intercept of 18.2' "Towards."

Since the LHA, at 348, is greater than 180°, Zn = Z, according to the rules, and is 151°. The plot is carried out in the regular way, stepping-off the intercept from the Assumed Longitude and Latitudes used in the solution. Except for the "rounding off" which is done in the process and creates small differences, the line of position should be virtually the same as that derived in the Vol. I solution.

All this doesn't sound too difficult, you say, so why all the disclaimers about star sights. The answer lies in the practice of taking, timing and recording the observation, especially if you're trying to do it by yourself—an operation akin to that of the one-armed paper hanger with the hives. As you try it, I think you'll see why I suggest there is a wide open area for your ingenuity.

First, of course, you have to identify the star you are going to shoot—although this can be done afterwards if the situation absolutely demands it. The identification is accomplished by estimating roughly the time you will be making the observation. Sometimes you can approximate this by the time you took your sights the day before, or you can make a calculation of the time of twilight from the sun's rising and setting data in the almanac. Either way, for your first experience get there early.

TABLE 5.—Correction to Tabulated Altitude

2	43 44 45	46 47 48	49 50 51	52 53 54	55 56 57	58 59 60	$\frac{d}{'}$
)	0 0 0	0 0 0	0 0 0	0 0 0	0 0 0	0 0 0	0
	1 1 1	1 1 1	1 1 1	1 1 1	1 1 1	1 1 1	1
	1 1 2	2 2 2	2 2 2	2 2 2	2 2 2	2 2 2	2
?	2 2 2	2 2 2	2 2 3	3 3 3	3 3 3	3 3 3	3
3	3 3 3	3 3 3	3 3 3	3 4 4	4 4 4	4 4 4	4
1	4 4 4	4 4 4	4 4 4	4 4 4	5 5 5	5 5 5	5
4	4 4 4	5 5 5	5 5 5	5 5 5	6 6 6	6 6 6	6
5	5 5 5	5 5 6	6 6 6	6 6 6	6 7 7	7 7 7	7
6	6 6 6	6 6 6	7 7 7	7 7 7	7 7 8	8 8 8	8
6	6 7 7	7 7 7	7 8 8	8 8 8	8 8 9	9 9 9	9
7	7 7 8	8 8 8	8 8 8	9 9 9	9 9 10	10 10 10	10
3	8 8 8	8 9 9	9 9 9	10 10 10	10 10 10	11 11 11	11
3	9 9 9	9 9 10	10 10 10	10 11 11	11 11 11	12 12 12	12
)	9 10 10	10 10 10	11 11 11	11 11 12	12 12 12	13 13 13	13
)	10 10 10	11 11 11	11 12 12	12 12 13	13 13 13	14 14 14	14
)	11 11 11	12 12 12	12 12 13	13 13 14	14 14 14	14 15 15	15
	11 12 12	12 13 13	13 13 14	14 14 14	15 15 15	15 16 16	16
2	12 12 13	13 13 14	14 14 14	15 15 15	16 16 16	16 17 17	17
3	13 13 14	14 14 14	15 15 15	16 16 16	16 17 17	17 18 18	18
3	14 14 14	15 15 15	16 16 16	16 17 17	17 18 18	18 19 19	19
4	14 15 15	15 16 16	16 17 17	17 18 18	18 19 19	19 20 20	20
5	15 15 16	16 16 17	17 18 18	18 19 19	19 20 20	20 21 21	21
5	16 16 16	17 17 18	18 18 19	19 19 20	20 21 21	21 22 22	22
6	16 17 17	18 18 18	19 19 20	20 20 21	21 21 22	22 23 23	23
7	17 18 18	18 19 19	20 20 20	21 21 22	22 22 23	23 24 24	24
3	18 18 19	19 20 20	20 21 21	22 22 22	23 23 24	24 25 25	25
3	19 19 20	20 20 21	21 22 22	23 23 23	24 24 25	25 26 26	26
)	19 20 20	21 21 22	22 22 23	23 24 24	25 25 26	26 27 27	27
)	20 ? ?	? ? ??	23 23 24	24 ?? ??	?? ?? ??	?? 28 28	28

Figure 9–6. Excerpt from Table 5, Pub. No. 249, showing Correction to Tabulated Altitude for d of 55, and Incremental Minutes, 18

With your estimated time of observation translated to GMT, you can take the GHA of Aries from the almanac as you do in working a star sight. By applying your rough estimate of longitude to it, you have an approximation of LHA♈. Now you can go one of two easy ways. You can enter Vol. I of Pub. No. 249 with your estimated latitude and, opposite the LHA♈ you have just worked out, read out the seven selected stars and their altitudes and azimuths. This is probably the simplest and quickest way, and is especially to be recommended if you intend to work your sights with Vol. I.

Alternatively, you can acquire a nifty device, originally developed for the Navy and called the Rude Star Finder. It is now published commercially and is very useful. You choose one of its plastic templates for your nearest 10° increment of latitude, and orient it over the base to your estimated LHA of Aries. The altitudes and azimuths of your visible stars are read off by inspection.

There are other methods, of course, although none are as simple as these two. In any event, the careful navigator will have prepared a list of the approximate locations of the stars he might shoot to take with him on deck, for then the fun begins.

The procedure is simple. You observe the star's altitude by centering its image on the line of the horizon, recording the exact time and sextant reading, and then move on to the next star. On a small boat, the trick is how you do this by yourself with only two hands. Some of the ideas I've tried are discussed in Chapter 12, Practical Wrinkles. The principal problem is that you have a narrow band of time in which to make your observations, the period between the time the sky is too bright to see the stars, and the time it is too dark to see the horizon. The finer optics of the more expensive sextants, as well as their larger fields of view, extend this effective time, but it's brief at best.

You can find your star in the sextant's field of view by presetting the instrument to the star's estimated altitude from your prepared list, and panning along the horizon under it. I prefer to hold the sextant upside down with my left hand and aim it directly at the star. Then, having located the star in the eyepiece, I move the index arm with my right hand to bring the horizon *up* to the star. This done, the sextant is reversed, the image is found to be practically on the horizon, and I proceed with the fine adjustment to complete the sight.

If you are fortunate enough to have a capable and interested watchstander on deck to help you, your task will be made immeasurably easier. He can read off the estimated altitudes and azimuths on your prepared list of stars, and can take and record the time of the sight as you call out, "Mark!" Your night vision won't be spoiled by having to use a light, and you can concentrate on getting the best possible results with your sextant. Even under these happy circumstances, however, I go back to what I said in Chapter 1 and reconfirm that, under the normal conditions in a small boat at sea, the sun is a much more constant companion to the short-handed navigator.

10. Special Cases

As you further your study of celestial, you may want to explore some of the navigator's "special situations" such as "Latitude by Noon Sun", also called "Latitude by Meridian Altitude," and the less frequently used "Latitude by Polaris".

The "noon sight" is a tradition among mariners. The old-timers used the simple and direct reduction of an observation of the sun at Local Apparent Noon (LAN) to obtain a quick and reliable latitude. They so depended upon it that it became common practice on long passages simply to run down to the latitude of the destination, at a point well to seaward of it, and then turn east or west as called for until a landfall was made, using noon sights all the way. Even today, the noon sight is standard practice in the Navy and Merchant Marine.

The particular joy of the noon sight is that the sextant altitude of the sun at Local Apparent Noon changes very slowly, so that the demand for precision in time is greatly lessened. Then, too, the calculations to be made afterward are child's play. Without going into the geometry of the situation, let me give you a few simple directions to try.

In preparing to take a noon sight, the navigator needs to know the approximate time of Local Apparent Noon —that moment when the sun crosses the observer's meridian (longitude), and is at its maximum altitude for the day. This is known as the "Meridian Passage" and is the time when the sun's GHA is exactly equal to your west longitude (or 360° minus your east longitude). The *Nautical Almanac* gives the time of Meridian Passage for each day at your "standard" meridian, which is the one nearest your longitude of the meridians spaced at each 15° increment east or west of Greenwich (0°). Figure 10-1 shows the almanac table which gives the time

Figure 10–1. Auxiliary Table from Nautical Almanac Daily Pages showing time of Sun's Meridian Passage

Day	SUN		Mer. Pass.
	Eqn. of Time		
	00ʰ	12ʰ	
	m s	m s	h m
9	01 00	00 55	11 59
10	00 49	00 43	11 59
11	00 37	00 31	11 59

of the sun's Meridian Passage for each of the three days shown on that page.

It is necessary to correct the standard meridian time for the amount your estimated longitude is east or west of it at the approximate time of LAN, adding or subtracting respectively, and expressing that difference in terms of time. Each degree of longitude is equal to four minutes, and each minute of longitude is equal to four seconds of time. If the Meridian Passage of the sun occurs at your standard meridian at 1159 on June 9, and your estimated position at that time will be 4½° of longitude, or 18 minutes of time, *east* of it, LAN can be expected to occur at 1141 local time.

In practice, the navigator starts shooting several minutes before his predicted time of LAN, and follows the sun up by a series of observations as its altitude increases. As the sun appears to "hang", just before the readings start to decrease, the navigator records the maximum altitude attained and calculates his latitude directly by the following steps:

· Correct the sextant altitude in exactly the same manner as for a normal sun sight, obtaining Ho.

· Note the sun's bearing from you, north or south.

· Subtract the Ho from 90°, arriving at an intermediate figure, "Zenith Distance" (z), which is marked with the name *opposite* the sun's bearing.

· Take the sun's declination at your estimated time of LAN—expressed in GMT—from the almanac, add it to z if the same name, take the difference if opposite name.

· The answer will be the observer's latitude, with the name—North or South—of the larger of z or declination.

Let's try a practical example so you can see how easy this special case is to work. A workform is shown which will help you follow the action. On June 9th, 1978, you estimate your longitude about the time of LAN to be 70° 30′ West. You observe the sun at a maximum sextant altitude of 71-58.8. Your Index Correction is −1.4 and your height of eye is nine feet.

Using the table in Figure 10-1, you have determined that the Mer Pass of the sun on June 9th takes place at 1159 at your nearest "standard" meridian, which is 75° W., or 4½° west of your estimated longitude. Therefore, the sun will arrive at your position earlier by 18 minutes (4½° times 4 minutes per degree), or 1141. This would be 1241, if you were keeping Daylight Saving Time, and is 1641 GMT.

An alternate way to estimate the time of LAN is to enter the SUN column of the almanac's daily pages (Figure 10-3) with your estimated west longitude (360° minus the longitude if east). Find the sun's GHA next lower than your longitude and subtract it. With that difference, enter the yellow Increments and Corrections tables in the back of the almanac, and find in the SUN column, the degrees and minutes of difference you calculated in subtracting the GHA from your longitude. The time of LAN in GMT will then be the *hour* opposite the GHA you identified, and the *minutes and seconds* corresponding to the angular difference with which you entered the Increment table.

After you have made your LAN observation, you cor-

rect the hs in the usual way to obtain Ho, using the
regular SUN table inside the front cover of the almanac,
an excerpt of which is shown in Figure 10-2. Applying

*Figure 10–2. Portion of Altitude Correction Tables from
Nautical Almanac showing Dip correction (−2.9' for
Ht. of Eye of 9 ft.) and Sun's Lower Limb Refraction
correction (+15.6' for App. Alt. of 71° 54.9') for observations April–September*

A2 ALTITUDE CORRECTION TABLES 10°-90°

OCT.—MAR. SUN APR.—SEPT.				DIP			
App. Alt.	Lower Limb	Upper Limb	App. Alt. Lower Limb	Upper Limb	Ht. of Eye	Corrⁿ	Ht. of Eye	Corrⁿ Ht. of Eye Corrⁿ

OCT.—MAR. SUN			APR.—SEPT.		DIP				
App. Alt.	Lower Limb	Upper Limb	App. Alt. Lower Limb	Upper Limb	Ht. of Eye	Corrⁿ	Ht. of Eye	Ht. of Eye	Corrⁿ
° '			° '		m		ft.	m	
9 34	+10·8	−21·5	9 39	+10·6 −21·2	2·4	−2·8	8·0	1·0	−1·8
9 45	+10·9	−21·4	9 51	+10·7 −21·1	2·6		8·6	1·5	−2·2
9 56	+11·0	−21·3	10 03	+10·8 −21·0	2·8	−2·9	9·2	2·0	−2·5
10 08	+11·1	−21·2	10 15	+10·0	3·0	−3·0	9·8	2·5	−2·8
						−2·1			
54 49	+15·5 +15·6	−16·7	57 02	15·3 −16·5 +15·4 −16·4	18·4	−7·5	60·5		
59 23	+15·7	−16·6	61 51	+15·5 −16·3	18·8	−7·6	62·1	130	−11·1
64 30	+15·8	16·5	67 17	+15·6 −16·2	19·3	−7·7	63·8	135	−11·3
70 12	+15·9	16·4	73 16	+15·7 16·1	19·8	−7·8	65·4	140	−11·5
76 26	+16·0	16·3	79 43	+15·8 16·0	20·4	−7·9	67·1	145	−11·7
83 05	+16·1	16·2	86 32	+15·9 15·9	20·9	−8·0	68·8	150	−11·9
90 00			90 00		21·4	−8·1	70·5	155	12·1

App. Alt.   Apparent altitude

Sextant altitude corrected for index error and dip.

your IC of −1.4, the Dip of −2.9, and the R correction of +15.6, your Ho works out to 72-10.1. You noted that the sun bore south of you at the time of your observation, so the Ho was marked accordingly.

Subtracting your Ho from 90°, you next obtain the Zenith Distance (z), in your example 17-49.9, which is marked *North,* the opposite of the sun's bearing. The declination is found in the SUN column of the daily pages of the almanac, a portion of which is shown in Figure 10-3.

Figure 10–3. Excerpt from Nautical Almanac showing Declination data for Sun on June 9 at 16ʰ and 17ʰ G.M.T.

1978 JUNE 9, 10, 11 (FRI., SAT., SUN.)

G.M.T.	SUN			MOON			
	G.H.A.	Dec.		G.H.A.	v	Dec.	d
d h	° ′	° ′		° ′	′	° ′	′
9 00	180 15.0	N22 52.8		142 58.0	12.9	N16 50.0	3.9
01	195 14.9	53.1		157 29.9	12.9	16 46.1	4.0
02	210 14.8	53.3		172 01.8	12.9	16 42.1	4.1
03	225 14.7	·· 53.5		186 33.7	13.0	16 38.0	4.2
04		·· 7		201 05.7	13 0	··	
	v 13.6 Nec.						
A 13	15 13.5	55.6		331 54.7	13.2	15 53.1	4.9
Y 14	30 13.3	55.8		346 26.6	13.2	15 48.2	5.0
15	45 13.2	·· 56.0		0 58.8	13.2	15 43.2	5.0
16	60 13.1	56.2		15 31.0	13.3	15 38.2	5.1
17	75 13.0	56.5		30 03.3	13.3	15 33.1	5.2
18	90 12.8	N22 56.7		44 35.6	13.3	N15 27.9	5.2
19	105 12.7	56.9		59 07.9	13.3	15 22.7	5.
20	120 12.6	57.1		73 40.2	13.3	15 17.4	5.
21	135 12.5	·· 57.3		88 12.5	13.4	15 12.1	5
22	150 12.4	57.5		102 44.9	13.4	15 06.6	5
23	165 12.2	57 7				15 01 1	5

DATE	June 9
Est.λ	70-30 W
Std. Mer.	75 W
Corr in time	13m
Mer. Pass.	1159 EST / 1259 EDST
LAN	1241 EDST
GMT	1641
hs	71 - 58.8
IC	- 1.4
D	- 2.9
ha	71 -54.5
R	+15.6
Ho	72 - 10.1 S
90°	90 - 00.0
-Ho	72 - 10.1 S
z	17 - 49.9 N
Dec	22 - 56.4 N
L	40 - 46.3
(name)	North

Workform for Latitude by Sun at Local Apparent Noon

In your example, you will remember the GMT of LAN was estimated to be 1641, so, by eye, you interpolate between the Dec readings for 1600 and 1700, arriving at a Dec of 22-56.4 N, the *same* name as z. Because the names are the same, Dec and z are *added,* and the total is your latitude, 40° 46.3' North.

Since the sun's GHA is equal to your west longitude (or 360° minus your east longitude) at LAN, it figures that, given the *exact* time of LAN, you could look up the GHA for that moment and make a direct determination of your longitude. The problem, however, is obtaining a sufficiently exact time of LAN. You will find in practicing your noon sights that the sun appears to "hang" for several moments at its maximum altitude, and it is difficult to pick the exact second when the maximum occurs. To overcome this, some practitioners take a series of sights, starting 15 minutes or more before the estimated time of LAN, recording or plotting the times and altitudes as the sun moves up. Then, as the readings start to decrease, an attempt is made to take the exact time of each matching altitude and, by averaging the times of common altitudes, an estimate of the actual time of LAN can, theoretically, be derived.

This procedure requires a lot of time and trouble and, if clouds, sea conditions or ship's movement during the series introduces an error in several of the sights, the results are likely to be quite uncertain. On the other hand, with accurate time available, single sun lines, which can be taken quickly and at any time convenient to the navigator, will produce better results with less effort. Using the technique for a "running fix," reviewed in Chapter 12, a good noon position can be determined simply and accurately. This is undoubtedly why that technique has found favor in the Navy and Merchant Marine, and with most knowledgeable yachtsmen.

Another "special situation" is a Latitude by Polaris. With a passing nod to the general problem of taking star sights from a small boat, discussed in Chapter 9, an altitude of the star Polaris—the "Pole Star"—also permits a direct solution for latitude. The procedure is thought to have been among the earliest exercises in nautical astronomy, though you'll not often use it because of available modern alternatives. The steps are brief and simple:

- Correct the sextant altitude in exactly the same manner as for a normal *star* sight, obtaining Ho.

- Obtain LHA♈ by applying your estimated longitude to GHA♈ which is taken from the almanac.

- Enter the Polaris Tables at the back of the *Nautical Almanac* (Figure 10-7), noting the value of a_0 for your LHA♈ in the upper section of the table.

- Descend the same column and note the value of a_1 opposite your approximate latitude.

- Descend to the third level in the same column and note the value of a_2 opposite the appropriate month.

- Then, Ho $+ a_0 + a_1 + a_2 - 1° =$ Latitude. Since the pole star is visible only in north latitudes, it goes without saying that the latitude by Polaris is always North.

You will have noticed in the second step that the *estimated,* rather than an *assumed* longitude is to be applied to the GHA♈. It is important to use your best estimate of the longitude at the time of the sight and calculate the LHA♈ to minutes of arc, since that is a determinate in the accuracy of the latitude derived.

Let's demonstrate the Latitude by Polaris technique with a practical example. The figures have been entered in the workform which you can follow step by step. On June 9th, 1978, in DR position Latitude 40° 51′ North, Longitude 69° 44′ West, you observe Polaris at morning twilight; GMT 08-57-30. The sextant altitude is 41-11.7, the Index Correction is − 1.4 and your height of eye is nine feet.

By now you are quite familiar with correcting the sextant altitude to obtain Ho. The Polaris observation is treated just like a normal star and uses the same table inside the front cover of the almanac (Figure 10-4). Your IC of − 1.4′, your Dip correction of − 2.9′, and your R correction of − 1.1′ are applied in the usual manner to your hs to obtain the ha and then the Ho which, for your example, will be 41-06.3

Next, you take out the GHA of Aries from the almanac's daily pages (Figure 10-5) for the hour of GMT (08ʰ) which is 17-20.3.

This tabular GHA has to be adjusted for the incremental 57 minutes and 30 seconds of GMT which is done in the regular way in the Increments and Corrections tables in the back of the almanac (Figure 10-6). In your example, the increment will be 14-24.9 which, when added to the tabular GHA, gives you a GHA♈ of 31-45.2.

To complete the second step of the procedure, it is necessary to apply the longitude to the GHA♈ to obtain

ALTITUDE CORRECTION TABLES 10°-90°

STARS AND PLANETS			DIP			
App. Alt. Corr^n		App. Additional Alt. Corr^n	Ht. of Eye Corr^n		Ht. of Eye	Ht. of Eye Corr^n
° ′ ′		1978	m		ft.	m ′
9 56	−5·3	VENUS	2·4	−2·8	8·0	1·0 − 1·8
10 08	−5·2	Jan. 1-July 20	2·6	−2·9	8·6	1·5 − 2·2
10 20	−5·1	° ′	2·8	−3·0	9·2	2·0 − 2·5
10 33	−5·0	42 ┼ 0·1	3·0	−3·1	9·8	2·5 − 2·8
10 46			3·2			3·0 − 3·0

33 40	−1·4	VENUS	14·7	−6·8	48·4	95 − 9·5
37 48	−1·3	Jan. 1-Mar. 22	15·1	−6·9	49·8	
40 08	−1·2	° ′	15·5	−7·0	51·3	100 9·7
42 44	−1·1	41 ┼ 0·2	16·0	−7·1	52·8	105 − 9·9
45 36	−1·0	75 ┼ 0·1	16·5	−7·2	54·3	110 − 10·2
48 47	0·9	Mar. 23-Dec. 31	16·9	−7·3	55·8	115 •10·4
52 18	0·8	° ′	17·4	−7·4	57·4	120 10·6
56 11	0·7	60 ┼ 0·1	17·9	−7·5	58·9	125 10·8
60 28	0·6		18·4	−7·6	60·5	
65 08	−0·5		18·8	−7·7	62·1	130 11·1
70 11	−0·4		19·3	−7·8	63·8	135 11·3
75 34	−0·3		19·8	−7·9	65·4	140 11·5
81 13	0·2		20·4	−8·0	67·1	145 11·7
87 03	0·1		20·9	−8·1	68·8	150 11·9
90 00	0·0		21·4		70·5	155 12·1

App. Alt.   Apparent altitude

Sextant altitude corrected for index error and dip.

Figure 10–4. Portion of Altitude Correction Tables from Nautical Almanac showing Dip correction (−2.9′ for Ht. of Eye of 9 ft.) and Polaris' Refraction correction (−1.1′ for App. Alt. of 41° 07.4′)

1978 JUNE 9, 10, 11 (FRI., SAT., SUN.)

G.M.T.	ARIES	VENUS −3.4		MARS +1.5		J(
	G.H.A.	G.H.A.	Dec.	G.H.A.	Dec.	(
d h	° ′	° ′	° ′	° ′	° ′	
9 00	257 00.5	143 33.0	N23 35.7	107 07.1	N13 39.8	15
01	272 03.0	158 32.3	35.3	122 08.3	39.3	17
02	287 05.5	173 31.6	34.9	137 09.4	38.8	18
03	302 07.9	188 30.8	·· 34.4	152 10.6	·· 38.3	2C
04	317 10.4	203 30.1	34.0	167 11.8	37.8	2]
05	332 12.9	218 29.4	33.5	182 13.0	37.3	2:
06	347 15.3	233 28.6	N23 33.1	197 14.2	N13 36.7	2(
07	2 17.8	248 27.9	32.7	212 15.3	36.2	2(
08	17 20.3	263 27.2	32.2	227 16.5	35.7	27
F 09	32 22.7	278 26.4	·· 31.8	242 17.7	·· 35.2	29
R 10	47 25.2	293 25.7	31.3	257 18.9	34.7	30
I 11	62 27.7	308 25.0	30.9	272 20.0	34.2	32
D 12	77 30.1	323 24.2	N23 30.4	287 21.2	N13 33.7	33
A 13	92 32.6	338 23.5	30.0	302 22.4	33.2	35
Y 14	107 35.0	353 22.8	29.5	317 23.6	32.7	
15	122 37.5	8 22.0	·· 29.0	332 24.8	·· 32.2	2
16	137 40.0	23 21.3	28.6	347 25.9	31.7	3
17	152 42.4	38 20.6	ⁿ⁹˙¹	2 27.1	31.2	5

Figure 10–5. Excerpt from the Nautical Almanac show-
ing G.H.A. of Aries on June 9 at 08ʰ

LHA♈ which is needed to enter the Polaris Table. You will recall that we are to use our best estimated longi- tude for accuracy, and that LHA♈ will be figured to the minute. In our example, notice that we have a case where the west longitude, 69° 44.0′, is greater than the GHA from which it is to be subtracted. Accordingly, we have added 360° to GHA and proceeded with the calcula- tion, arriving at an LHA♈ of 322-01.2. This figure is then

INCREMENTS AND CORRECTIONS 57ᵐ

57ᵐ	SUN PLANETS	ARIES	MOON	v or d	Corrⁿ	v or d	Corrⁿ	v or d	Corrⁿ
s	° ′	° ′	° ′	′	′	′	′	′	′
00	14 15·0	14 17·3	13 36·1	0·0	0·0	6·0	5·8	12·0	11·5
01	14 15·3	14 17·6	13 36·3	0·1	0·1	6·1	5·8	12·1	11·6
02	14 15·5	14 17·8	13 36·5	0·2	0·2	6·2	5·9	12·2	11·7
03	14 15·8	14 18·1	13 36·8	0·3	0·3	6·3	6·0	12·3	11·8
04	14 16·0		37·0	0·4	0·4				
		14 24·1					8·3	14·7	14·1
28	14 22·0	14 24·4	13 42·7	2·8	2·7	8·8	8·4	14·8	14·2
29	14 22·3	14 24·6	13 43·0	2·9	2·8	8·9	8·5	14·9	14·3
30	14 22·5	14 24·9	13 43·2	3·0	2·9	9·0	8·6	15·0	14·4
31	14 22·8	14 25·1	13 43·4	3·1	3·0	9·1	8·7	15·1	14·5
32	14 23·0	14 25·4	13 43·7	3·2	3·1	9·2	8·8	15·2	14·6
33	14 23·3	14 25·6	13 43·9	3·3	3·2	9·3	8·9	15·3	14·7
34	14 23·5	14 25·9	13 44·2	3·4	3·3	9·4	9·0	15·4	14·8
35	14 23·8	14 26·1	13 44·4	3·5	3·4	9·5	9·1	14·9

Figure 10–6. Excerpt from Nautical Almanac Increments and Corrections Tables showing Aries increment for 57ᵐ 30ˢ

used to enter the Polaris (Pole Star) Tables which are found just before the yellow pages in the back of the *Nautical Almanac.* Figure 10-7 shows the portion of this table applicable to your example.

With the LHA♈, enter the column in which the heading includes your LHA and, opposite the nearest degree of LHA in the left hand column, extract the a₀ correction, interpolating if necessary by eye. For your LHA♈ of

POLARIS (POLE STAR) TABLES, 1978

L.H.A. ARIES	300°– 309°	310°– 319°	320°– 329°	330°– 339°	340°– 349°	350°– 359°
	a_0	a_0	a_0	a_0	a_0	a_0
°	° ′	° ′	° ′	° ′	° ′	° ′
0	1 01·6	0 52·9	0 44·4	0 36·3	0 28·8	0 22·3
1	1 00·8	52·1	43·5	35·5	28·1	21·7
2	0 59·9	51·2	42·7	34·7	27·4	21·1
3	59·0	50·3	41·9	33·9	26·8	20·5
4	58·2	49·5	41·1	33·2	26·1	20·0
5	0 57·3	0 48·6	0 40·3	0 32·4	0 25·4	0 19·4
6	56·4	47·8	39·4	31·7	24·8	18·9
7	55·5	46·9	38·6	31·0	24·1	18·4
8	54·7	46·1	37·8	30·2	23·5	17·9
9	53·8	45·2	37·0	29·5	22·9	17·4
10	0 52·9	0 44·4	0 36·3	0 28·8	0 22·3	0 16·9

Lat.	a_1	a_1	a_1	a_1	a_1	a_1
°	′	′	′	′	′	′
0	0·2	0·2	0·2	0·3	0·4	0·4
10	·2	·2	·3	·3	·4	·5
20	·3	·3	·3	·4	·4	·5
30	·4	·4	·4	·4	·5	·5
40	0·5	0·5	0·5	0·5	0·5	0·6
45	·5	·5	·5	·6	·6	·6
50	·6	·6	·6	·6	·6	·6
55	·7	·7	·7	·7	·6	·6
60	·8	·8	·8	·7	·7	·7
62	0·9	0·8	0·8	0·8	0·7	0·7
64	0·9	0·9	·9	·8	·8	·7
66	1·0	1·0	0·9	·9	·8	·7
68	1·1	1·0	1·0	0·9	0·9	0·8

Month	a_2	a_2	a_2	a_2	a_2	a_2
	′	′	′	′	′	′
Jan.	0·6	0·6	0·6	0·6	0·7	0·7
Feb.	·4	·5	·5	·5	·5	·6
Mar.	·3	·3	·3	·4	·4	·4
Apr.	0·2	0·2	0·2	0·2	0·3	0·3
May	·3	·2	·2	·2	·2	·2
June	·4	·3	·3	·2	·2	·2
July	0·5	0·5	0·4	0·3	0·3	0·3
Aug.		·2	·6			

Latitude = Apparent altitude (corrected for refraction)

$$-1° + a_0 + a_1 + a_2$$

322-01.2, the reading for 322° of 42.7' did not require adjustment.

Descending the same column, in a similar fashion, you extract the a_1 value, which is 0.5, opposite your approximate latitude (41°). Again no interpolation was needed in your example.

In the third level of the table opposite your month of observation (June), you take out of the same column with which you have been working, the a_2 value, which is 0.3. The final step is to add a_0, a_1, and a_2 to your Ho of 41-06.3, subtracting one whole degree according to the rules, and arriving at your latitude, which is 40° 49.8' North.

You might have wondered about using Table 6, Correction Q for Polaris, in the back of Vol. I of Pub. No. 249 as an alternate method. If you did, you would have proceeded in the same way through the second step to obtain LHA♈, extracted the single value Q from Table 6 for your LHA♈, and applied it to your Ho to get latitude. Although it saves a step or two, this method is not as accurate as the almanac method and I would not recommend it unless you don't have access to a *Nautical Almanac.*

Figure 10–7. Portion of Polaris Tables from Nautical Almanac showing the three-step correction for L.H.A. Aries 322° 01.2' (42.7'), for Latitude 41° (0.5'), and for June (0.3'). Note formula at bottom to determine Latitude

NAVIGATOR'S WORKBOOK

DATE	June 9
hs	41 - 11.7
IC	.- 1.4
D	-2.9
ha	41-07.4
R	-1.1
Ho	41-06.3
W	08-57-30
corr	00
GMT	08 - 57-30
gha ♈	17-20.3
incr	14 - 24.9
GHA♈	31-45.2
	(+360)
	391-45.2
aλ	69-44.0
LHA♈	322-01.2
aL	41 N
Ho	41-06.3
-1°	40-06.3
+a₀	42.7
+a₁	0.5
+a₂	0.3
L	40-49.8 N

Workform for Latitude by Polaris

11. A Little Theory

You were promised at the outset a minimum of theory and a maximum of practicability, and I intend to keep that promise. But, if you are one of those who are more comfortable with a basic feel for the principles involved, let's digress for a moment and, with a little license to simplify some of the technicalities, look at the concept underlying the sights you have been taking. If you find it confusing rather than enlightening, simply move on or, if by now you're becoming hooked on celestial, you may want to dig still deeper and the classic texts are there and waiting.

What you have actually done is to measure the altitude of the selected body by sextant observation which enabled determining the distance in arc from your actual position to the point on the earth's surface directly beneath the body which is called the "Geographical Position" or GP. Next, you computed the theoretical arc distance of the GP from an Assumed Position (AP) nearby your actual, though unknown, position. This computation (for Hc) was done, of course, with the sight reduction tables by inspection, but the tables simply performed a solution by spherical trigonometry of the so-called "Navigational Triangle".

Fig. 11-1 shows a typical navigational triangle projected on the earth's surface. One side of the triangle is the arc distance from the Assumed Position (AP) to the adjacent, or "elevated" pole. It amounts to 90° minus the

Figure 11–1. The Navigational Triangle projected on Earth

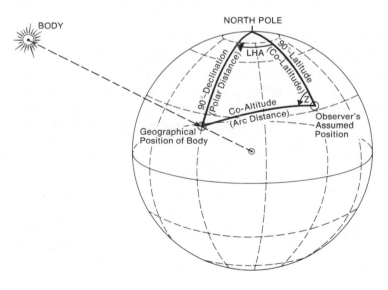

Assumed Latitude, which is called "Co-Latitude". The second side of the triangle is the arc distance from the Geographical Position (GP) to the pole, and this is often referred to as the "Polar Distance". It amounts to 90° minus the body's declination. The angle formed at the pole between the meridians of the AP and GP is the Local Hour Angle. Recalling that, with the help of the almanac, you had obtained a value for the LHA and Dec of the body observed, and had selected a latitude for your Assumed Position, you have in hand the information necessary to quantify those two sides and the included angle of the Navigational Triangle, which is all you require to solve for the remaining side and angle you want for your computation.

By applying the formulae of spherical trig, you proceed to determine the third side, which represents the arc distance from the AP to the GP, identified by navigators as the "Co-Altitude" or "Zenith Distance", and the adjacent angle, the azimuth angle Z.

As the next-to-last step in determining your celestial line of position, you will recall that you compared Ho with your computed Hc—in principle the arc distance from the AP to the GP was compared to the arc distance from the *actual* position to the GP. Using the difference in arc distance as the "intercept" (a), you plotted the line of position from a point at the appropriate bearing and distance from the Assumed Position. Fig 11–2 might help dispel the fog surrounding the foregoing.

In the sketch you will note the GP of the body on the earth's surface directly beneath it. The angle of altitude at the observer's position Ho, and the corresponding angle at the Assumed Position, Hc, is indicated along with the circles of equal altitude—the locus of all points

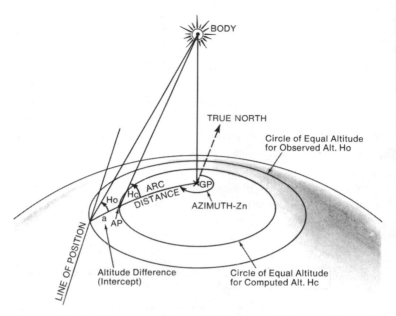

Figure 11–2. Construction of Circles of Equal Altitude illustrating the use of the Intercept in establishing a Line of Position. Since Hc is greater than Ho, the Intercept is away from the direction of the azimuth.

on earth where the altitude of the body will be the same at the time of observation. The difference in the radii of these two circles (i.e. the difference in arc distance from the observer's position and the Assumed Position to the GP) is the Altitude Difference, or Intercept (a). With knowledge of the true Azimuth from your calculations, and whether Ho or Hc is greater, you have the compass direction of the line through the AP to the GP, and the information needed to apply the intercept toward or away from the GP. You will notice that the resulting line of position in the sketch is shown as a tangent at the point of intersection of the azimuth line and the circumference of the observer's equal-altitude circle. As explained earlier, the circle of equal altitude on the globe is generally so large that the short length of this position line is, to all intents and purposes, identical with the circumference.

12. Practical Wrinkles

Now that you have had a chance to see what celestial navigation is all about—at least as it affects the practicing small-boat navigator—it is time to admit you to the "back room" to view a few of the wrinkles that experienced navigators sometimes use to simplify their lives a little. As you gain experience, you'll add a number of tricks of your own, and that is truly part of the fun of celestial.

One particularly useful wrinkle is to do as much work as possible *before* going on deck to take a sight. This might include getting out the almanac and the appropriate volume of the sight reduction tables, checking the watch for the correct time and, in the case of star sights, working out a list of the approximate altitudes and azimuths of those you intend to shoot. This is also a good time to run up your DR on your chart or plotting sheet so you will be prepared to select an appropriate Assumed Latitude and Longitude for your sight reduction.

It is not a bad idea to consider, at this moment, where you are going to locate yourself to observe the body you want, and possibly to pop your head out for a moment to check the index error of your sextant. If you are planning on taking visual or electronic lines to cross with your celestial line, now is a good time to prepare for them too.

On deck in a small boat, finding a good place to position yourself while taking sights is always a challenge. With the sails up, it's often necessary to have several stations in mind in order to cover the whole horizon without being blanketed by the sails or interfered with by the rigging. In rough weather, it is desirable to be as high as possible without sacrificing your stability or safety. This is both to avoid a false horizon from nearby seas and for the very practical reason of keeping the sextant's optics from becoming clouded with salt spray. You will want to time your sights under these conditions to take them as your boat rises to the top of a wave, since your distant and correct horizon will be the tops of the waves out there.

While complete overcast may make it impossible to obtain a workable sight, one shouldn't give up in merely cloudy, hazy or partially obscured conditions. By lowering your height of eye as much as possible, you bring your horizon closer—often close enough to get adequate definition in haze. By proper use of the sextant's shades, the sun's image may be seen sufficiently well to bring one good limb to the horizon.

In broken clouds, a common condition, the navigator must be on deck and "at the ready" for that instant when a body peeks out from behind a cloud and he can get a quick shot. The three key words under trying conditions are height-of-eye, shades and patience—the latter the test of a real navigator.

If you have to take your own time, and need both hands, you might try hooking yourself to the standing rigging with a safety harness. Then concetrate on the details of sight-taking instead of fighting the motion of the boat.

If it is necessary to spend much time on deck in heavy weather, before or between sights, I've found it useful to protect the sextant with a light, plastic bag. I used to use a lanyard attached to the sextant too, and many navigators swear by this practice. It was just one more thing to get in the way, and I *never* let go of my sextant, so I don't use a lanyard save under exceptional conditions. Another good practice is to clean off the sextant's optics after they have been exposed to salt spray.

In Chapter 1 we mentioned a method to make certain that the sextant was absolutely vertical at the moment of measuring the altitude. This technique, known as "rocking" the sextant, consists of rotating it back and forth around the line of sight to the horizon so the body appears to swing in an arc as shown in Figure 12-1. A good stunt is to rock the sextant when the body is just above the point of tangency and then, when you are certain the body is at the bottom of its arc, lower it quickly to the horizon and take the reading. The arc becomes flatter as the altitude increases so that rocking

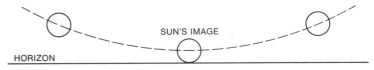

Figure 12–1. Path of the Sun's image when "rocking" the Sextant

with high altitude shots, such as the LAN sun, takes a bit of practice.

While taking a sun line, under appropriate conditions you may also have the opportunity to take a good bearing of it by your steering compass. The azimuth you derive from working your sight will be good enough to compare with that bearing to provide a valuable check on your compass. The difference between the true bearing and the compass bearing, corrected for local variation, is the ship's deviation on the particular heading—very useful to know at sea.

Rather than expose a valuable watch on deck, I like to use a cheap but readable one, which indicates seconds and is set to GMT. I check it against an accurate timepiece on board, which is, in turn, compared with radio time ticks each day. By holding the watch in my left palm, using my fingers to manipulate the tangent screw of the sextant, I have only to open my hand and glance up at the instant I'm satisfied with the sight. I do use a lanyard on the watch, which is attached to my wrist.

We have discussed some of the problems and techniques for shooting stars in Chapter 9, and this applies also to the planets. At both morning and evening twilight, the eastern horizon stars can be shot to advantage first—in the evening the east darkens first while in the morning it becomes visible first. With all the heavenly bodies, but especially the stars, it is desirable, when you have a completely free choice, to favor sights in the middle band of altitudes, say between 15° and 70°. The problem with low-altitude sights is that the refraction correction is much more critical, as a glance at the almanac will confirm, and the reduced light transmitted through the oblique angle to the earth's atmosphere makes for much fainter images. With the higher altitudes, and especially as the body approaches the zenith, it becomes increasingly difficult to be certain that the sextant is aimed correctly in azimuth and has been held absolutely vertically. This cautionary note should apply only when you have a totally free choice. The modern data will permit you to get at least passable results at the extremes of the altitude range when you have no better alternative.

On the subject of extremes, the altitude correction tables in the almanac have been established for average conditions of temperature (50° F.) and barometric pressure (29.83 inches). Only in the case of major deviations from these norms, and then only for very-low altitude observations do you need to be concerned. Table A4 at the front of the *Nautical Almanac* will provide the additional correction. It's seldom if ever applied in yachting weather.

Many navigators seem to prefer morning stars to sights at evening twilight. I suppose the reason for this is that it is often calmer in the early morning and, because you start with a sky full of stars, the identification problem may be a little easier. Then there is the advantage of going below to work out your sights in daylight rather than in a darkened ship. With my disclaimer on star sights in general, I think there are some arguments favoring the morning proponents, but, if you select and plan your round of stars in advance, I don't think the differences are extreme. Incidentally, a "quick and dirty" method of approximating the LHAϒ for star identification is to enter the ARIES table in the daily pages of the almanac using your *local* time for GMT, reading out the corresponding GHAϒ as LHAϒ. You will see that the answer is only as accurate as your proximity to the standard meridian of your time zone, and the trick is only simple in west longitudes.

You will recall the technique you learned in piloting for advancing a line of position and for producing a "running fix". For a refresher, Bowditch has a complete treatment of the subject. An extremely useful wrinkle with the sun is to take sights within an hour or two of LAN, and advance the resulting position lines to LAN to provide a three-line running noon fix. Figure 12-2 illustrates such an operation. The absolute accuracy of your running fix, of course, will be a function of the correctness of your estimate of your vessel's movement during the elapsed period. Therefore, the small-boat navigator strikes a balance between a short interval between the sights, so the estimated ship's movement will be more precise, and a longer period which allows the sun's azimuth to change more and, thus, give a better angle of intersection at the fix.

Other very useful sun lines—and this could apply to any other body as well—are those which you take when the sun's bearing is nearly ahead or astern, which are

POSITION PLOTTING SHEET

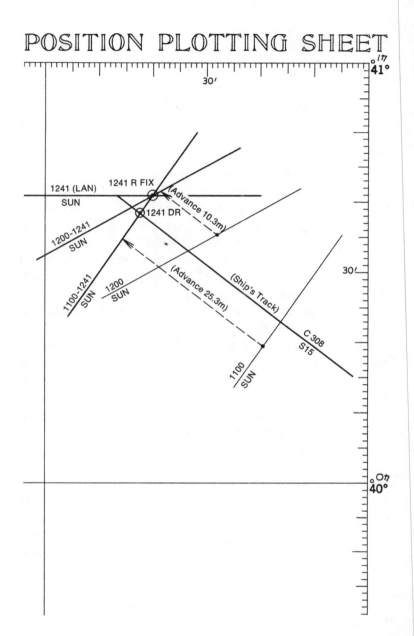

called "speed lines" (the 1100 line in Figure 12-2 is an example), or when the bearing is abeam, called a "course line". These single position lines can provide valuable information, short of a fix, as to your progress in direction or distance travelled.

There are two signs I consider to be the mark of a proficient navigator. One is a neat, carefully labelled plot, and the other is a well organized workbook. There are many choices in style, and as a guide to help you determine yours, as well as to epitomize the techniques presented in this text, there are included sample workforms for all the sights we have discussed. I have seen forms such as these combined into one, universal form, although it is slightly more complicated as a result; forms prepared in pads, or in a looseleaf format; and even strip forms designed to be used as a bookmark and positioned in the margin of the page of a blank workbook. The choice, as well as the opportunity to make further improvements to suit your own style, is your own.

As a final word, please don't overlook the explanation sections in the back of the almanac and in the front of the sight reduction tables which may help to extricate you from any depths into which I have lured you, but have failed to plumb to your satisfaction. As my old Navy Chief used to say, "When all else fails, try reading the directions".

How does one conclude an introduction to the small-boat navigator's world of celestial navigation? "Good

Figure 12–2. Plot of Running Fix advanced to the time of Local Apparent Noon

Luck" is inappropriate since it's your acquired knowledge and skill, and not my luck, that is going to keep you out of trouble. "Clear skies and calm seas" has a fine ring, but I'm afraid it's an unrealistic, though fond wish.

Perhaps my thoughts are best expressed by John Hamilton Moore, Bowditch's predecessor, who, in the preface to the *New Practical Navigator* in 1796 wrote, "My grand object has been to be concise yet comprehensive, explanatory in my definitions, perspicuous in my rules and examples, and, in a word, most carefully attentive to every particular that can further the acquisition of an art that has been the object of my pursuits throughout my life". May it enrich yours!

Glossary

Almanac—A publication containing the astronomical data required for the practice of celestial navigation, arranged by calendar date and time interval. The *Nautical Almanac,* recommended in this text, supplies the information from which the Greenwich Hour Angle and Declination of the principal celestial bodies can be determined for any instant of time.

Altitude Difference (a)—The difference between the Observed Altitude (Ho) and the Computed Altitude (Hc); commonly called the Intercept.

Apparent Altitude (App. Alt. or ha)—The Sextant Altitude (hs) corrected for index error and dip.

Arc Distance—The distance measured along a curve; in celestial navigation usually a portion of a great circle.

Assumed Latitude (aL)—The latitude at which the observer is assumed to be for the purpose of calculating the Computed Altitude (Hc) of a celestial body. Usually selected to the nearest whole degree.

Assumed Longitude (aλ)—The longitude at which the observer is assumed to be for the purpose of calculating the Computed Altitude (Hc) of a celestial body. Usually selected so that the Local Hour Angle (LHA) works out to a whole degree.

Assumed Position (AP)—The position assumed for calculating the Computed Altitude (Hc) of a celestial body, and the point from which the Altitude Difference, or Intercept (a) is plotted.

Azimuth (Z and Zn)—The uncorrected Azimuth (Z), also called Azimuth Angle, is measured from North (0°) or South (180°), clockwise or counterclockwise, through 180 degrees. The corrected, or True Azimuth (Zn) is measured from North (0°) clockwise through 360 degrees.

Celestial Sphere—An imaginary sphere, concentric with the earth and with the earth at its center, on which all the celestial bodies are presumed to be projected.

Co-altitude—90° minus the altitude; also called the Zenith Distance (z).

Co-latitude—90° minus the latitude.

Computed Altitude (Hc)—The altitude of a celestial

body at a given time and position as determined by computation.

Critical Table—A table in which a single value is tabulated for limiting increments of entry values as, for example, the almanac's altitude correction tables.

Dead Reckoning (DR)—A position derived by applying courses and distances sailed from the last known position.

Declination (Dec or dec)—The angular distance north or south of the celestial equator, corresponding to latitude on earth. The abbreviation *d* is used in the almanac to indicate the hourly change in declination.

Dip (D)—The angle between the true horizontal and the observer's line of sight to the visible horizon.

First Point of Aries (♈)—The point at which the sun's path intersects the celestial equator as it changes from south to north declination at the Vernal Equinox. Values for the Greenwich Hour Angle of Aries are tabulated in the daily pages of the almanac as if it were a celestial body, and the positions of all the stars are measured westward from that point by their Sidereal Hour Angles.

Fix—A position, determined without reference to a previous position, usually resulting from the intersection of two or more lines of position.

Geographical Position (GP)—The point on earth directly beneath a celestial body.

Great Circle—The circle formed by the intersection of a plane passing through the center of a sphere with the surface of the sphere.

Greenwich Hour Angle (GHA)—The angular distance measured westward from the meridian of Greenwich (0°) on the celestial sphere. Corresponds to longitude on earth. The small case abbreviation (gha) is often used to identify the uncorrected tabular value extracted from the GHA tables in the almanac.

Greenwich Mean Time (GMT)—Local mean time at the meridian of Greenwich (0°). Time signals, broadcast as "Coordinated Universal Time" (UTC), may vary by a fraction of a second from GMT as a result of irregular rotation of the earth. For practical purposes, the navigator uses the two times interchangeably.

Horizon Glass—The half-mirrored glass, attached to the frame of a sextant, through which the horizon is viewed.

Horizon Shades—The darkened glass which can be moved into place to reduce the intensity of light passing through the clear portion of the horizon glass.

Horizontal Parallax (H.P.)—The difference in altitude between that measured from the observer's position on the surface of the earth and that measured from the center of the earth. Of primary interest to the navigator in correcting altitudes of the moon, where the value is of significance because of the relative closeness of the moon to the earth, and an additional correction for it must be taken from the almanac.

Index Arm—The movable arm of a sextant.

Index Correction (IC)—The value which must be applied to correct the index error (failure to read exactly zero when the true and reflected images are in coincidence) of a sextant. Usually confirmed before or after each series of observations.

Index Shades—The darkened glass which can be moved into place between the mirror on the index arm and the eyepiece to reduce the intensity of the reflected image of a celestial body. It is essential that the reflection of the sun's image be reduced in intensity by use of the Index Shades as, otherwise, injury can result to the unprotected eye.

Inspection Tables—A volume of tabulated solutions from which an answer can be extracted by simple inspection.

Intercept (a)—The difference between the Observed Altitude (Ho) and the Computed Altitude (Hc).

Interpolation—The process of determining intermediate values between given, tabular values.

Line of Position—"A line on some point of which a vessel may by presumed to be located as a result of observation or measurement"—Bowditch.

Local Apparent Noon (LAN)—That moment when the sun crosses the observer's meridian and is at its maximum altitude for the day.

Local Hour Angle (LHA)—The angular distance measured westward from the observer's meridian on the celestial sphere.

Main Arc—That part of a sextant upon which the readings in degrees are inscribed. Sometimes called "the limb."

Meridian—A great circle through the geographical poles of the earth or celestial sphere. The meridian of Greenwich (0°) is called the prime meridian.

Meridian Passage (Mer Pass)—The time of meridian passage; when a celestial body crosses a given meridian.

Micrometer Drum—A device for making precise, small measurements on a sextant. The mechanism is referred to as an endless tangent screw, and a sextant so-equipped as an E.T.S. sextant.

Navigational Triangle—The spherical triangle whose points are the elevated pole, the celestial body, and the zenith of the observer projected on earth, which is solved in determining Computed Altitude (Hc) and Azimuth.

Noon Sight—Observation of the sun's altitude at Local Apparent Noon.

Observed Altitude (Ho)—The Apparent Altitude (ha) corrected for refraction (R), and, in the case of the moon, additionally for Horizontal Parallax (H.P.).

Polar Distance—The angular distance from the celestial pole; in celestial navigation, specifically, 90° minus the declination.

Polaris—The Pole Star, located less than 1° from the North Celestial Pole in the constellation Ursa Minor. Useful for a special-case latitude determination.

Refraction (R)—The correction due to the bending of light rays passing obliquely through the earth's atmosphere. The refraction correction in the *Nautical Almanac* includes, for the convenience of a single entry solution, other corrections such as semi-diameter, etc.

Semi-diameter (S.D.)—The angular distance from the center of a celestial body of finite diameter (e.g. the sun) to its outer edge or limb.

Sextant Altitude (hs)—The uncorrected altitude of a celestial body as measured directly by sextant observation.

Sidereal Hour Angle (SHA)—The angular distance measured westward on the celestial sphere, from the First Point of Aries through 360 degrees.

Sight Reduction Tables—Tables for solving the navigational triangle for Computed Altitude (Hc) and Azimuth (Zn). Pub. No. 249, *Sight Reduction Tables for Air Navi-*

gation, is used in this text.

Standard Meridian—A central meridian selected for a time zone, located at each multiple of 15° longitude east or west of Greenwich (0°).

Tabulated Altitude (Tab Hc)—The uncorrected value of Hc as extracted from the sight reduction table.

Variable Correction (*v*)—Small, additional corrections due to excesses of actual movement over the constant rates used in the body of the Increments and Corrections tables in the almanac.

Vernier—A scale for precise, small readings on a sextant.

Watch Time (W)—The time registered on the navigator's watch or clock.

Zenith—That point on the celestial sphere directly over the observer.

Zenith Distance (z)—The angular distance from the zenith; in celestial navigation, 90° minus the altitude, or Co-altitude.